Great White Sharks
in United States
Museums

Great White Sharks in United States Museums

ALESSANDRO DE MADDALENA
and WALTER HEIM

Foreword by SEAN R. VAN SOMMERAN

McFarland & Company, Inc., Publishers
Jefferson, North Carolina, and London

All drawings © 2009 Alessandro De Maddalena

LIBRARY OF CONGRESS CATALOGUING-IN-PUBLICATION DATA

De Maddalena, Alessandro, 1970–
Great white sharks in United States museums / Alessandro De Maddalena and Walter Heim ; foreword by Sean R. Van Sommeran.
    p.   cm.
Includes bibliographical references and index.

ISBN 978-0-7864-4183-9
softcover : 50# alkaline paper ∞

1. White shark—Collection and preservation—United States. 2. Zoological specimens—United States. 3. Zoological museums—United States—Catalogs. I. Heim, Walter (Walter D.) II. Title.
QL638.95.L3D39   2009   597.3'307473—dc22   2009036807

British Library cataloguing data are available

©2009 Alessandro De Maddalena and Walter Heim. All rights reserved

No part of this book may be reproduced or transmitted in any form or by any means, electronic or mechanical, including photocopying or recording, or by any information storage and retrieval system, without permission in writing from the publisher.

On the cover: *focal* shark jaws and teeth (photograph courtesy Bone Clones, Canoga Park, California); *from bottom left* shark "swimming" in the atrium, San Diego Natural History Museum (photograph courtesy Tim Murray); replica of shark (final taxidermy by Peter Wilson; photograph courtesy J. Evan Ward, Department of Marine Sciences of the University of Connecticut, Groton); captured shark preserved at the Museum of Comparative Zoology, Harvard University (photograph courtesy Andrew Williston); background ©2009 Shutterstock

Manufactured in the United States of America

McFarland & Company, Inc., Publishers
Box 611, Jefferson, North Carolina 28640
www.mcfarlandpub.com

To my wife, Alessandra, and my son, Antonio.
—Alessandro

To my wife, Beverly, who tolerates my many days
on the water chasing sharks.
—Walter

Contents

Acknowledgments	viii
Foreword by Sean R. Van Sommeran	1
Preface	3
1 — Biology, Ethology and Ecology of the Great White Shark	7
2 — Preserving and Reconstructing Great White Sharks	85
3 — Great White Shark Materials in Museums	113

    Alaska   113   North Carolina   156
    California   114   Pennsylvania   159
    Connecticut   139   South Carolina   161
    Florida   141   Texas   162
    Illinois   146   Virginia   162
    Massachusetts   148 Washington (State)   164
    New York   154   Washington, D.C.   165

4 — Great White Sharks in Aquariums (A Chronological Report)	168
Appendix I — Institutions Contacted	187
Appendix II — Reporting Specimens of Great White Sharks	195
Bibliography	199
Index	207

Acknowledgments

We must pay special homage to Beverly Heim (San Diego, California) who took the time to read and edit the entire manuscript. Very special thanks to David C. Powell for his continuous support, his endless kindness and his generosity in sharing his personal experiences, knowledge and material. Many thanks to respected colleague, great white shark expert, and dear friend Sean R. Van Sommeran for taking his precious time to read the manuscript and write the Foreword.

We thank the following people, as well as the institutions where they work, for freely sharing their observations and for their assistance in assembling material for this book: Nicola Allegri (Italy), Kurt Auffenberg (Florida Museum of Natural History, Gainesville, Florida), Donna Baron (Alaska State Museum, Juneau, Alaska), Larry Bell (Museum of Science, Boston, Massachusetts), Valerie Bell (South Florida Museum, Bradenton Florida), Jeff Benca (University of Washington, School of Aquatic & Fishery Sciences, Seattle, Washington), Rachel M. Berquist (Digital Fish Library, UCSD Center for Scientific Computation in Imaging, La Jolla, California), Chris Black (Middleton, Tasmania, Australia), Bone Clones (Canoga Park, California), Martha Bringhurst Bruno (Morehead City, North Carolina), Dan Brooks (Houston Museum of Natural Science, Houston, Texas), Barbara A. Brown (American Museum of Natural History, New York City, New York), Gregor M. Cailliet (Pacific Shark Research Center, Moss Landing Marine Laboratories, Moss Landing, California), Rosemary J. Carlton (Sheldon Jackson Museum, Sitka, Alaska), Daniel P. Cartamil (Scripps Institution of Oceanography, University of California San Diego, La Jolla, California), Julie Carter (National Ocean Service Marine Forensics Branch, Center for Coastal Environmental Health and Biomolecular Research, Charleston, South Carolina), José I. Castro (Mote Marine Laboratory, Sarasota, Florida), David Catania (California Academy of Sciences, San Francisco, California), James Chang (North Carolina State Museum of Natural Sciences, Raleigh, North Carolina), Stuart Chase (Berkshire Museum, Pittsfield, Massachusetts), David Cicimurri (Campbell Geology Museum, Clemson University, Clemson, South Carolina), Eugenie Clark (Mote Marine Laboratory, Sarasota, Florida), Ralph S. Collier (Shark Research Committee, Van Nuys, California), Cheri F. Collins (Connecticut Archaeology Center, Connecticut State Museum of Natural History, Storrs, Connecticut), Chris Conroy (Museum of Vertebrate Zoology, University of California, Berkeley, California), Charles F. Cotton (Department of Fisheries Science, Virginia Institute of Marine Science, College of William and Mary, Gloucester Point, Virginia), Gerald L. Crow (Waikiki Aquarium, University of Hawaii, Honolulu, Hawaii), Gianluca Cugini (Mediterranean Shark Research Group, Milan, Italy), Stephen Daratsos, Jr. (Los Angeles, California), Jon Dodrill (Florida Fish and Wildlife Conservation Commission, Tallahassee, Florida), Alison Eichelberger (North Museum of Natural History & Science, Lancaster, Pennsylvania), Richard Ellis (American Museum of Natural History, New York City, New York), Karren Elsbernd (Archives and Digital Collections, California Academy of Sciences, San Francisco, California), Lollo Enstad (San Diego Natural History Museum, San Diego, California), Manny Ezcurra (Monterey Bay Aquarium, Monterey, California), Jennifer Forrest (Museum of Discovery, Little Rock, Arkansas), Lawrence R. Frank (Center for Scientific Computation in Imaging, La Jolla, California), Callaghan Fritz-Cope (Pelagic Shark Research Foundation, Santa Cruz, California), Jeff Gage (Florida Museum of

Acknowledgments

Natural History, Gainesville, Florida), Chris Gannon (North Carolina State Museum of Natural Sciences, Raleigh, North Carolina), Suzanne M. Gendron (Ocean Park, Hong Kong, People's Republic of China), Olivier Glaizot (Musée cantonal de Zoologie, Lausanne, Switzerland), Jerry Goldsmith (Borrego Springs, California), Francesco Guerrazzi (Firenze, Italy), Jessica Heim (San Diego, California), Roger C. Helm (Division Environmental Quality, U.S. Fish and Wildlife Service, Arlington, Virginia), John C. Hewitt (Audubon Aquarium of the Americas, New Orleans, Louisiana), Eric J. Hilton (Department of Fisheries Science, Virginia Institute of Marine Science, College of William and Mary, Gloucester Point, Virginia), Susan Hochgraf (Biological Research Collections, University of Connecticut, Storrs, Connecticut), Ernst Hofinger (Hofinger Tier-Präparationen, Steyrermühl, Austria), Gabriela Hogue (North Carolina State Museum of Natural Sciences, Raleigh, North Carolina), Jeff Huber (Florida Museum of Natural History, Gainesville, Florida), Robert E. Hueter (Mote Marine Laboratory, Sarasota, Florida), Richard C. Hulbert, Jr. (Division of Vertebrate Paleontology, Florida Museum of Natural History, University of Florida, Gainesville, Florida), Debra A. Ingrao (Mote Marine Laboratory, Sarasota, Florida), Karen Jeffries (Monterey Bay Aquarium, Monterey, California), Scott Jervas (Berkshire Museum, Pittsfield, Massachusetts), Kevin M. Johns (United States Department of Agriculture, Tampa, Florida), Ray Keyes (Aquatic Design Systems Inc., San Diego, California), Carolyn Kirdahy (Museum of Science, Boston, Massachusetts), Leah Kissel (Waikiki Aquarium, University of Hawaii, Honolulu, Hawaii), Michel Krafft (Musée cantonal de Zoologie, Lausanne, Switzerland), Scott LaGreca (Berkshire Museum, Pittsfield, Massachusetts), Vivian Lam (Swire Institute of Marine Science, University of Hong Kong, Hong Kong, People's Republic of China), You-Hung Lin (Taipei, Taiwan, Republic of China), Chris Lowe (CSULB Sharklab, California State University, Long Beach, California), Kyle Luckenbill (Academy of Natural Sciences, Philadelphia, Pennsylvania), John G. Lundberg (Department of Ichthyology, Academy of Natural Sciences, Philadelphia, Pennsylvania), Katherine Pearson Maslenikov (University of Washington, School of Aquatic & Fishery Sciences, Seattle, Washington), Barbara Mathe (American Museum of Natural History Library, New York City, New York), John E. McCosker (California Academy of Sciences, San Francisco, California), Craig Moritz (Museum of Vertebrate Zoology, University of California, Berkeley, California), Tim Murray (San Diego Natural History Museum, San Diego, California), Paige Newman (Sea World San Antonio, San Antonio, Texas), David Nolan (Studio Y Creations, Calgary, Alberta, Canada), Northeast Fisheries Science Center (Woods Hole, Massachusetts), Mike O'Haver (Lyons and O'Haver, La Mesa California), John O'Sullivan (Monterey Bay Aquarium, Monterey, California), Debbie Paselk (Humboldt State University Natural History Museum, Arcata, California), Ken Peterson (Monterey Bay Aquarium, Monterey, California), David C. Powell (Pacific Grove, California), Antonella Preti (Southwest Fisheries Science Center, National Marine Fisheries Service, San Diego, California), Patricia Punch Pearson (Atomic Props, St. Paul, Minnesota), Stephen C. Quinn (American Museum of Natural History, New York City, New York), Irv Quitmyer (Florida Museum of Natural History, University of Florida, Gainesville, Florida), John E. Randall (Bernice P. Bishop Museum, Honolulu, Hawaii), Kimberly Rawson (Berkshire Museum, Pittsfield, Massachusetts), Tom Reidarson (Sea World San Diego, San Diego, California), Marci Bynum Robertson (Museum of Discovery, Little Rock, Arkansas), Robert H. Robins (Florida Museum of Natural History, University of Florida, Gainesville, Florida), Mary Anne Rogers (The Field Museum, Chicago, Illinois), J. Romine (Department of Fisheries Science, Virginia Institute of Marine Science, College of William and Mary, Gloucester Point, Virginia), Pacifique Rugira (Sea World San Diego, San Diego, California), John Rupp (Point Defiance Zoo and Aquarium, Tacoma, Washington), Gaspare Schillaci (Milan, Italy), Tim W. Scott (Virginia Aquarium & Marine Science Center, Virginia Beach, Virginia), Natasha Seibel (Virginia Aquarium & Marine

Acknowledgments

Science Center, Virginia Beach, Virginia), Jeffrey A. Seigel (Natural History Museum of Los Angeles County, Los Angeles, California), Mike Shaw (San Diego, California), Nadine Simak (Mote Marine Laboratory, Sarasota, Florida), Domenico Sorrenti (Ganzirri, Italy), Jenny Slafkosky (Monterey Bay Aquarium, Monterey, California), Victor G. Springer (Smithsonian Institution National Museum of Natural History, Washington, D.C.), Wayne C. Starnes (North Carolina State Museum of Natural Sciences, Raleigh, North Carolina), John D. Stevens (CSIRO Division of Marine Research, Hobart, Tasmania, Australia), Charles Stevenson (Mote Marine Laboratory, Sarasota, Florida), Melanie Stiassny (American Museum of Natural History, New York City, New York), Leighton Taylor (Napa Valley, California), Patricia Turner (Apex Predators Program, National Marine Fisheries Service, Narragansett, Rhode Island), Sean R. Van Sommeran (Pelagic Shark Research Foundation, Santa Cruz, California), Dave Varley (Atomic Props, St. Paul, Minnesota), H.J. Walker, Jr. (Scripps Institution of Oceanography, University of California, San Diego, La Jolla, California), J. Evan Ward (Department of Marine Sciences, University of Connecticut, Groton, Connecticut), Mary Warrick (Florida Museum of Natural History, Gainesville, Florida), Catherine Weisel (Museum of Comparative Zoology, Harvard University, Cambridge, Massachusetts), Joost Wenderich (Keep Smiling Diving Divecenter, Reeuwijk, The Netherlands), Bill Wieger (Shepherdsville, Kentucky), Randy Wilder (Monterey Bay Aquarium, Monterey, California), Jeffrey T. Williams (Smithsonian Institution National Museum of Natural History, Washington, D.C.), Philip Willink (The Field Museum, Fish Division, Chicago, Illinois), Andrew Williston (Museum of Comparative Zoology, Harvard University, Cambridge, Massachusetts), Chuck Winkler (Southern California Marine Institute, San Pedro, California), Sabine Wintner (Natal Sharks Board, Umhlanga, KwaZulu-Natal, South Africa), Eric Zamora (Florida Museum of Natural History, Gainesville, Florida), Melissa L. Zielinski (Humboldt State University Natural History Museum, Arcata, California), Marco Zuffa (Museo Archeologico "Luigi Donini," Ozzano dell'Emilia, Italy), and Donald D. Zumwalt (MAR^{3}INE, San Pedro, California).

For their help, support and friendship, our sincere gratitude goes to Alessandra Baldi, Antonio De Maddalena, Pinuccia De Maddalena, Emilio De Maddalena, Eleonora De Maddalena, Elisabetta De Maddalena, Isabella De Maddalena, Sauro Baldi, the students of the BCM school, Francesco Guerrazzi, Gianfranco Della Rovere, Matteo Messa, Michele Masera, Antonella Preti, Massimo Albini, Gaspare Schillaci, Giorgio Martello Panno, Alessandro De Marinis, Ralph S. Collier, Sean R. Van Sommeran, Brian May, Paul Rodgers, the Mediterranean Shark Research Group and the Italian Ichthyological Society.

Foreword by Sean R. Van Sommeran

In keeping with the legacy of the great philosopher Aristotle, Alessandro De Maddalena has been gathering and organizing specimens, artifacts, historical and classical records, artworks and photographs regarding the natural history and accrued human interpretations and descriptions of sharks and has been updating the scientific records to our times.

Ever since the first white shark tooth was handed over to Aristotle by some ancient mariner, the realization and understanding of this marvelous and sometimes monstrous denizen of the oceans has been advanced by the discerning gaze and musings of Alessandro De Maddalena. He has created a renaissance out of current conventional wisdoms and historical interpretations of existing records and artifacts, which has at once verified and sometimes corrected long-standing beliefs, and has also brought light to some of the more obscure treasures, specimens and collections.

Since the description by Herodotus of the Lamia's wrath upon his foes at Athos in 492 B.C., sharks have been understood to be something ominous, forces of nature and, according to Leonidas of Tarentum, agents of Apollo and the gods. Despite being one of the most often depicted and easily recognized species of sharks, the white shark, *Carcharodon carcharias*, remains one of the most poorly understood and feared species of predator on our planet.

With the current work Alessandro De Maddalena brings the country's scattered artifacts and collections out of the doldrums of vague interpretation and onto the examination table of our modern times. In this effort he brings to bear his personal intellect and scientific discipline as well as those of extant and ancient mariners and a fleet of higher institutions.

By way of worldwide networking, careful consideration and study, this latest installment by De Maddalena brings the bulk of the collective white shark treasures from around the United States into sharp focus and context.

Sean R. Van Sommeran is executive director of the
Pelagic Shark Research Foundation, Santa Cruz, California.

Preface

Alessandro De Maddalena

This book is the first complete treatise on great white shark (*Carcharodon carcharias*) materials preserved in the museums and institutes of the United States of America. This is not just a scientific work, this is also a work of love. The reason for this apparently strange statement will be clear as soon as the reader finishes reading these opening lines.

Someone may ask himself why a book on this subject was written by an Italian. I think that the reply to this question is pretty intriguing and it is useful to understand why this book has been conceived (I like this word, since a book is almost like a baby to its author). I may try to reply to the question by saying that writing this book has been a consequence of the fact that I have dedicated over thirty years of my life to the study of sharks (thirteen as a profession) publishing over a hundred articles as well as thirteen other books on sharks. But this may not be sufficient explanation. The question of why an Italian author had to cover such a specific subject still remains, so I think I have to try to find a better and more articulated reply. So, let's try.

There are actually three possible good replies to such a question. The first is simply because a work of this kind did not exist before, so I felt that someone had to do it, and who cares in the end if the author is an Italian or an American researcher? It is true that both historically and at present, the United States of America has been at the leading edge of the study of sharks, and that this great nation, having given birth to the greatest number of shark specialists, has consequently produced the highest amount of quality scientific works in this field. So, one may wonder why such a specialized work on great white sharks preserved in United States institutions has not been written before. Really I don't know, but in the end, I decided to take the responsibility and the pleasure to fill this gap.

The second possible reply is that in these thirteen years of researching sharks in a professional context, starting from the preparation of my thesis at the University of Milan, the study of white shark museum materials has been one of the main interests and one of the main focuses of my activity. Having recently completed a long and deep study of great white shark materials preserved in European museums, the possibility of preparing a similar work based in the United States, the nation that has the greatest amount of materials of this species preserved in the whole world, seemed too tempting to let it escape. I think that this second explanation may completely satisfy the reader.

There is a third possible reply to the initial question, and it's a much more personal one. I started seriously studying sharks a little bit before I reached ten years of age. It was the late 70s, and the internet era was still far away in the future (things like these were vaguely hypothesized about only in science fiction novels). So, I built the basis of my knowledge on sharks by devouring books on sharks at home and reading with the deepest interest scientific papers on the same subject at the libraries of the Aquarium & Civic Hydrobiological Station, and at the Natural History Museum in Milan, my hometown, which 30 years ago was already a horrible cement grey town quite far from the sea. The names of authors that most impressed me at this early age for the quality of their publications were those of some of the greatest American ichthyologists and shark experts, including Henry B. Bigelow, William C. Schroeder, Henry W. Fowler, Perry W. Gilbert, John E. Randall, Victor G. Springer, Ralph S. Collier and Wilbur I. Follett. At that time, and in the decades that followed, their names remained strongly impressed in my memory, and the respect that I have for these researchers, veritable pioneers in our field of shark research, is still great. Today we have so much specialization in our field that it is really hard to find someone who has the deep knowledge and the real interest necessary to apply a multi-angle approach to the study of sharks like those pioneers did. Having almost reached 40 years of age, I've felt the need to pay homage to the persons who provided me the basis of my knowledge about the biology, ethology and ecology of sharks. This book is intended to be an homage to the above cited ichthyologists, the U.S. shark researchers without whom the work of all other shark researchers that have followed would not exist. We, the ichthyologists of the world, have all learned from them and must be grateful to them for their passion and devotion. This book is a way to thank them for the hard work they did during their respective careers. But this book is also a way to thank them for this little crazy idea they put in my heart and in my mind so many years ago and for the fact that for some reason that I'll never completely understand, it seems that they will never abandon me. The power of attraction that sharks exercise on me is still strong, like that of certain dreams that capture our hearts and keep them in the sphere of their almost magical influence.

People often ask me why I love sharks, and especially great white sharks. The truth is that I don't exactly know. The most sincere reply I can give to this question is that I love sharks because they're so immensely beautiful. They captured my heart over 30 years ago like the most beautiful of all women, my wife, did 15 years ago. This may sound absurd, but for me this is the only explanation that makes sense to explain my neverending passion for these ancient creatures. Cultivating this attraction for all these years has been like following a dream and trying to make it real. Well, I don't know how and why all this has happened, but today I'm still "*Working on a dream,*" like Bruce Springsteen loves to say...

Walter Heim

As a young boy, I grew up fascinated with fish and in particular, sharks. This fascination was not a good fit for a boy growing up in a desert town. As I grew older, I became interested in sport fishing and diving. Soon, every family vacation was spent in San Diego, California. As a fisherman, sharks were a nuisance and as a diver, they were to be feared. I remember seeing the original *Jaws* movie and diving the next day always looking over my shoulder. After high school, I studied marine biology in order to become a fish scientist, but that never happened. Over the years, I have been blessed with a good job that has provided for my family and allows me to play marine biologist on the weekend. Being fascinated with sharks, I pursued them as a game fish and put many in the freezer. As I became interested in underwater photography, I was able to apply my fishing skills to photographing and tagging sharks. My favorite subject is the mako shark, which is kin to the white shark. Being an engineer, I recognize that lamnid sharks are a remarkably designed and beautiful fish. My summers now revolve around shark expeditions and my winter activities include stocking the freezer with mackerel for my summer shark expeditions. My contribution to the field of shark science has been the many sharks I have tagged for the scientific community and fisheries managers, and the articles and books I have shared with Alessandro. It can be a lot of work, but I love it.

1

Biology, Ethology and Ecology of the Great White Shark

Classification

Sharks belong to the kingdom Animal, the phylum Chordata, the subphylum Vertebrata, the superclass Pisces, the class Chondrichthyes, the subclass Elasmobranchii, the superorder Selachimorpha. The class Chondrichthyes consists of sharks, rays and chimaeras, which are cartilaginous fish. Cartilage is a light and flexible connective tissue. It is composed of specialized cells called chondrocytes that produce a large amount of extracellular matrix composed of collagen fibers, elastin fibers, and an abundant ground substance rich in proteoglycan. Cartilage is found in many areas in the human body, including the articular surface of joints, the ear, the nose, and the airways of the respiratory system. Cartilaginous fish have skeletons composed of cartilage (the only bony tissues are found in their teeth and scales). In contrast, the class Osteichthyes or bony fish, have skeletons made primarily of bone, like those found in humans (see page 37, "Skeletal System").

The superorder Selachimorpha is divided into eight orders, and these are divided into 34 families that include 479 species of sharks (this number includes a few dubious species that may not be valid). The eight orders of sharks are: Hexanchiformes (frilled and cow sharks), Squaliformes (dogfish sharks), Pristiophoriformes (saw sharks), Squatiniformes (angel sharks), Heterodontiformes (bullhead sharks), Orectolobiformes (carpet sharks), Lamniformes (mackerel sharks) and Carcharhiniformes (ground sharks). The order Lamniformes includes seven families: Odontaspididae, Mitsukurinidae, Pseudocarchariidae, Megachasmidae, Alopiidae, Cetorhinidae, and Lamnidae. The family Lamnidae, also called mackerel sharks, includes three genera: *Carcharodon*, *Isurus* and *Lamna*. The family Lamnidae has five species: great white shark *Carcharodon carcharias* (Linnaeus, 1758), shortfin mako *Isurus oxyrinchus* (Rafinesque, 1809), longfin mako *Isurus paucus* (Guitart Manday, 1966), porbeagle *Lamna nasus* (Bonnaterre, 1788) and salmon shark *Lamna ditropis* (Hubbs and Follett, 1947) (Compagno, 1984). All five lamnid species are present in U.S. waters (Castro, 1983). The genus *Carcharodon* includes only one living species, *Carcharodon carcharias*, the great white shark.

Carl Linnaeus, a Swedish zoologist, botanist, ecologist and physician, laid the foundations for the modern scheme of scientific nomenclature. He gave the name *Squalus carcharias* to the great white shark in 1758. Today, the name of the great white

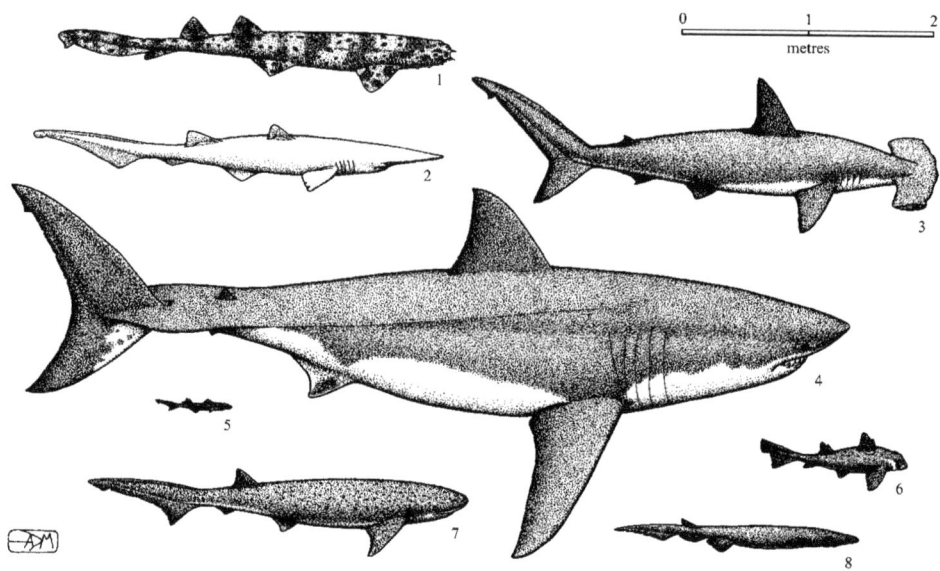

The size of some species of sharks compared to the size of the great white shark. (1) spotted wobbegong *Orectolobus maculatus* (Bonnaterre, 1788); (2) goblin shark *Mitsukurina owstoni* (Jordan, 1898); (3) scalloped hammerhead *Sphyrna lewini* (Griffith and Smith, 1834); (4) great white shark *Carcharodon carcharias* (Linnaeus, 1758); (5) Velvet belly *Etmopterus* spinax (Linnaeus, 1758); (6) horn shark *Heterodontus francisci* (Girard, 1854); (7) broadnose sevengill shark *Notorynchus cepedianus* (Peron, 1807); (8) frilled shark *Chlamydoselachus anguineus* (Garman, 1884).

shark accepted by the scientific community is *Carcharodon carcharias* (Linnaeus, 1758). The rules of taxonomy require that we pay homage to Linnaeus for having been the first to properly describe the great white shark, by citing his name after the shark's scientific name. The name Linnaeus is included in parentheses because the scientific name used today is not the same that was proposed by the Swedish author. In fact, the great white shark is classified within another genus, *Carcharodon*, which was proposed by Andrew Smith in an 1838 publication by Johannes Müller and Friedrich Henle. The Latin name of the great white shark genus is derived from the Greek *kàrkharos*, which means "serrated," and *odón*, which means "tooth," referring to the strongly serrated margins of the teeth. After Linnaeus described the great white shark, many other authors re-described the same species and gave it different names, which today are considered synonymous with *Carcharodon carcharias* and are not used. Among these synonyms are: *Carcharias lamia*, *Squalus lamia*, *Carcharias verus*, *Squalus lamia*, *Carcharias rondeletti*, *Squalus vulgaris*, *Carcharodon smithii*, *Carcharodon rondeletii*, *Carcharodon capensis*, *Carcharias atwoodi*, *Carcharias vorax*, *Carcharias maso*, *Carcharodon albimors* (Compagno, 1984).

1— Biology, Ethology and Ecology of the Great White Shark

Common Names

In the English language, sharks of the family Lamnidae are named mackerel sharks, because of some morphological features they share with bony fish belonging to the family Scombridae, which include the mackerels and tunas. These features include a hydrodynamic design based on a streamlined and spindle-shaped body, a conical snout, lateral keels on the sides of the caudal peduncle and the caudal fin, and a half-moon-shaped (lunate) caudal fin.

The name "great white shark," comes from its distinct white ventral surface. There are many secondary names in the English language, including white shark, white pointer, blue pointer, white death, and maneater (Compagno, 2001; Ellis and McCosker, 1991).

Among the different languages all over the world, some of the common names given to the great white shark are witdoodshaai (Afrikaans), peshkagen njeringrenes (Albanian), kalb (Arabic), bijela ajkula (Bosnian), bai sa, da bai sa (Cantonese), tauró blanc (Catalan), velika bijela psina, pas ljudožder (Croatian), žralok bílý (Czech), hvid Haj (Danish), witte haai (Dutch), wahsh (Egyptian), valkohai (Finnish), grand requin blanc (French), weißer hai (German), lefkos karkarias (Greek), qarha levana (Hebrew), squalo bianco, grande

The shortfin mako shark *Isurus oxyrinchus*. In the English language, sharks of the family Lamnidae are called mackerel sharks because of some morphological features they share with bony fishes belonging to the family Scombridae. These features include a hydrodynamic design based on a streamlined and spindle-shaped body, a conical snout, lateral keels on the sides of the caudal peduncle and a half-moon shaped caudal fin (courtesy Jessica Heim).

squalo bianco, pescecane (Italian), hohojirozame (Japanese), kelb abjad (Maltese), bai sha, da bai sha (Mandarin), kalb (Moroccan), hvithai (Norwegian), żarłacz biały (Polish), tubarâo branco (Portuguese), rechin alb, marele rechin alb (Rumanian), seldevaja akula (Russian), velika bijela ajkula (Serbian), žralok biely vel'ký (Slovak), beli morski volk (Slovenian), tiburón blanco, jaquetón blanco (Spanish), vithaj (Swedish), kelb el b'har (Tunisian), canavar köpekbaligi (Turkish) (Adem Hamzic, pers. comm., 2008; Barrull and Mate, 2002; De Maddalena and Baensch, 2008; De Maddalena & Hollà, 2006; Ellis and McCosker, 1991; Froese and Pauly, 2008; Zoran Kljajic, pers. comm., 2008; Vivian Lam, pers. comm. 2009; You-Hung Lin, pers. comm. 2009; Lipej *et al.*, 2004).

Evolution

While the cartilaginous skeleton of the shark rapidly disintegrates after death (complete skeletons are preserved only in very rare cases), teeth fossilize easily because they are highly calcified (see page 29, "Digestive System"). Shark teeth are among the most

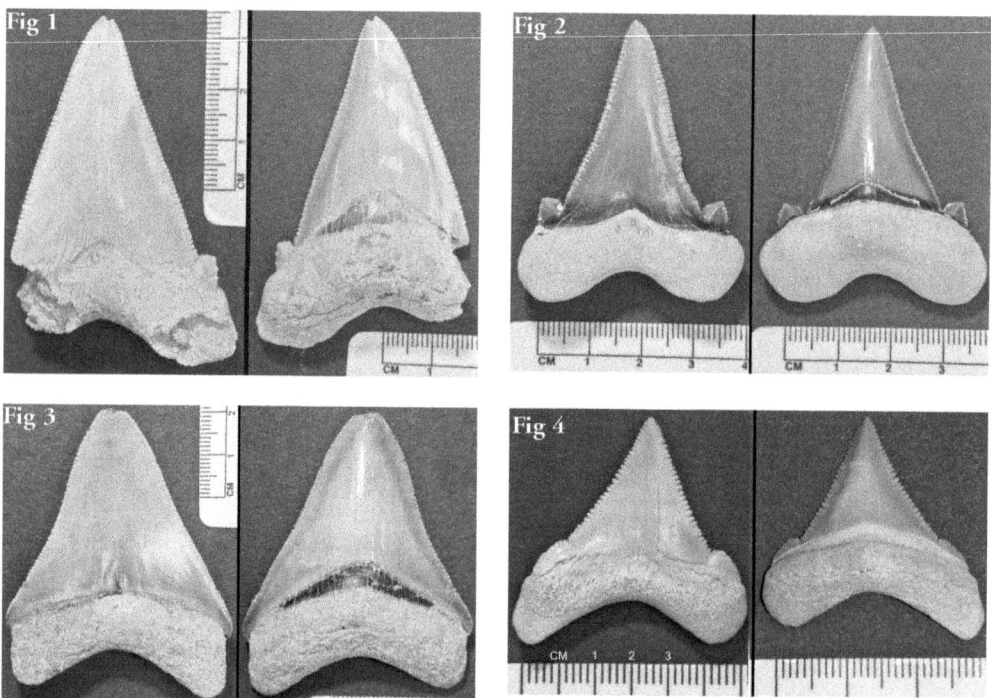

Fossil teeth of some extinct species of the genus *Carcharodon*, preserved at the Campbell Geology Museum, Clemson University, Clemson, South Carolina. Figure 1—*Carcharodon angustidens* (cat. no. BCGM 2770). Figure 2—*Carcharodon auriculatus* (cat. no. BCGM 5437). Figure 3—*Carcharodon megalodon* (cat. no. BCGM 4465). Figure 4—*Carcharodon subauriculatus* (without cat. no.) (courtesy David Cicimurri, Campbell Geology Museum, Clemson University, Clemson, South Carolina).

1—Biology, Ethology and Ecology of the Great White Shark

abundant of vertebrate fossils. Shark vertebrae and tiny dermal denticles (see page 25, "Skin") were occasionally preserved as fossils owing to their partial calcification.

Sharks arose some 425 million years ago, in the Silurian period (Martin, 1995). They evolved from the placoderms, a group of extinct armored bony fish (the earliest branch of the jawed fish). The retention of a similar body form by prehistoric and modern sharks suggests that early in their evolutionary history these fish developed morphological and anatomical characteristics that make them successful animals, well adapted to their environment (Long, 1995; Maisey, 1987).

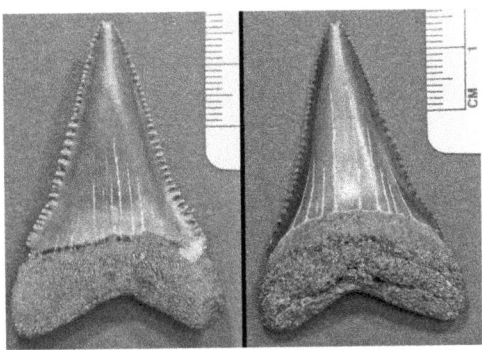

Fossil teeth of *Carcharodon carcharias*, preserved at the Campbell Geology Museum, Clemson University, Clemson, South Carolina, with cat. no. BCGM 2225 (courtesy David Cicimurri, Campbell Geology Museum, Clemson University, Clemson, South Carolina).

The earliest fossils in the genus *Carcharodon* have been found in the Paleocene (55.8 to 65.5 million years ago). *Carcharodon orientalis* (Sinzow, 1899) is the oldest species found to date. The extinct megatooth shark *Carcharodon megalodon* (Agassiz, 1835) attained an estimated 1590 cm total length and was the largest macropredator shark that has ever lived (Gottfried et al., 1996). Gottfried et al. (1996) inferred that, in comparison to the great white shark, the megatooth shark had a broader chondrocranium with a more domed cranial roof and a shorter rostrum, less elevated orbits, more massive jaws coupled with a more robust dentition, a higher vertebral count and proportionately larger fins.

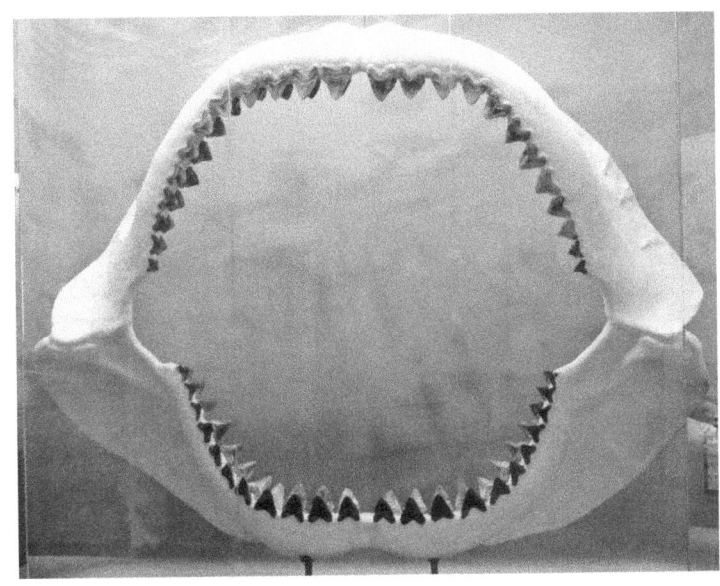

An approximately 350 cm wide by 270 cm tall reconstructed set of jaws of the extinct megatooth shark *Carcharodon megalodon*, on exhibit at the South Florida Museum in Bradenton, Florida. It was prepared by Cliff Jeremiah of Florida Paleontological Supply, Inc., Jacksonville, Florida (photograph courtesy South Florida Museum, Bradenton, Florida).

Carcharodon carcharias did not appear until the Late Miocene (11.6 to 5.3 million years ago) (Applegate and Espinosa-Arrubarrena, 1996). *Carcharodon carcharias* is the only living species of the genus *Carcharodon*.

External Anatomy

Great white sharks are easy to recognize. They are large in size, ranging in length from 1.2 to at least 7.0 meters. The caudal fin is lunate in shape with the caudal fin lower lobe almost as long as the upper lobe. The gill slits are long and the body is massive. The pectoral fins are long, and like all lamnid sharks, wide keels are present on the caudal peduncle. The mouth is wide and has large triangular teeth with serrated margins. The lower teeth are prominent and even visible when the shark has its mouth closed (De Maddalena and Baensch, 2008). The morphology of a great white shark is described in detail below.

Body form varies considerably among shark species, and is related to the shark's way of life. Pelagic sharks are usually much more hydrodynamic than benthic sharks ("pelagic" refers to all species living in the open sea, while "benthic" refers to species living most of the time on the sea floor). The great white shark is a pelagic species with a massive, spindle-shaped body.

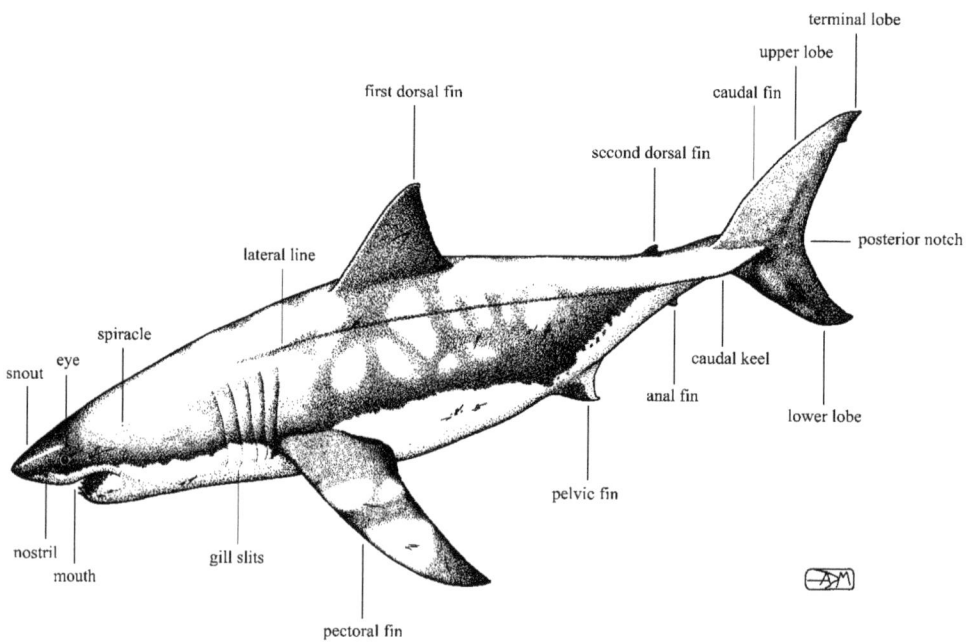

External anatomy of the great white shark.

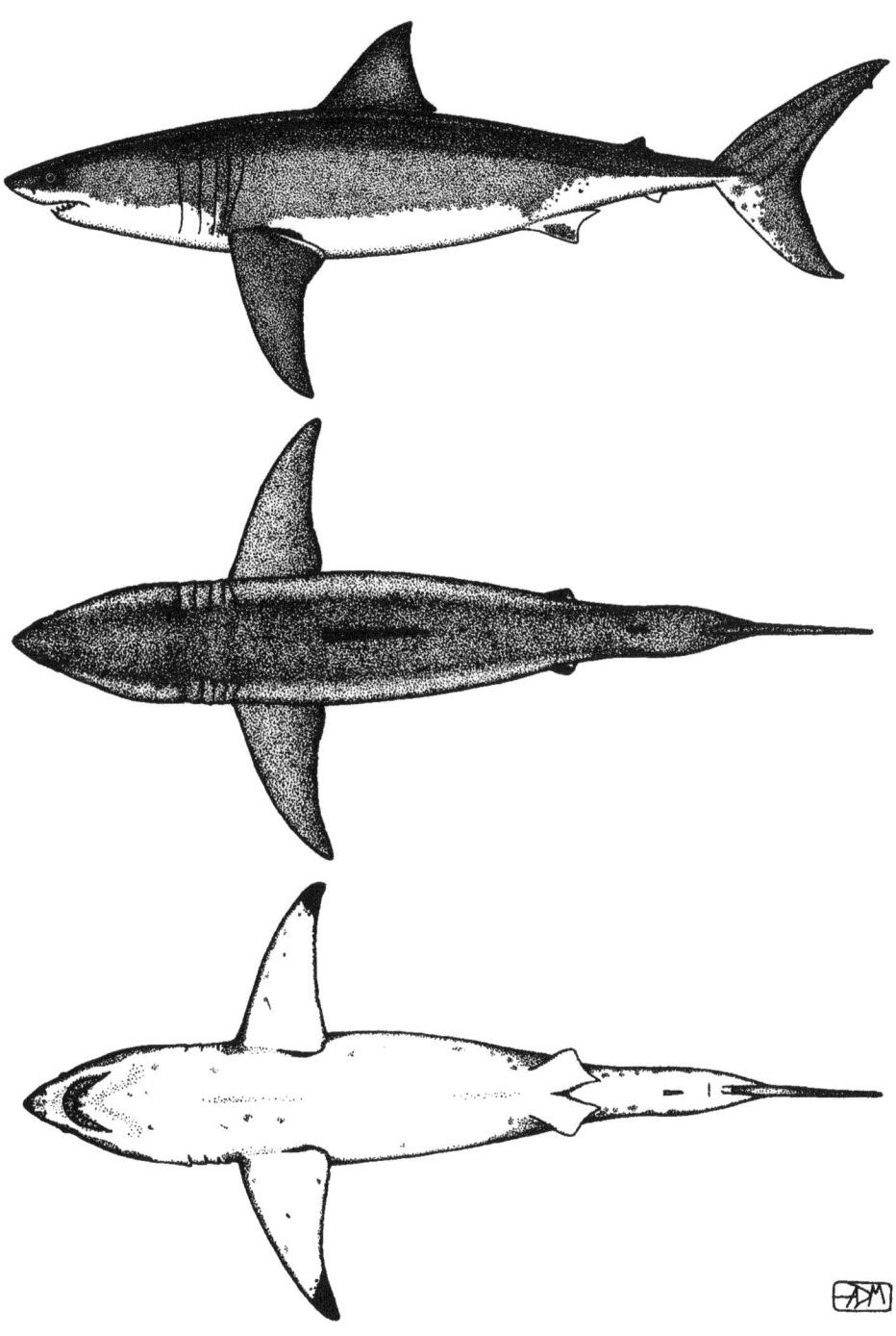

The great white shark as viewed from the lateral, dorsal and ventral sides.

The snout is large, conical and pointed. The eyes are relatively small, dark, and circular in shape. The eyes lack a nictitating membrane (see page 34, "Nervous System and Sensory Perception"). The nostrils are relatively small, located lateroventrally on the snout. The nostrils are partially covered by a nasal flap that separates sea water flowing into the nostril from water flowing out. There are very small spiracles (see page 26, "Respiratory System") located on the sides of the head, posterior to the eyes and anterior to the gill slits. The great white shark has five very long gill slits, all located anterior to the origin of the pectoral fins. The fifth gill slit is slightly more oblique than the others, which are parallel to each other.

Like most sharks, great white sharks have eight fins, including two pectoral, two pelvic, first dorsal, second dorsal, anal and caudal fins. Except for the caudle fin, each fin has a base, anterior and posterior margin, and a tip or apex. The posterior margin extends past the base and the lower corner of this extension is the free rear tip of the fin. The origin of a fin is the anterior-most point of the base and a fin insertion refers to the posterior-most point of the base. The caudal fin is lunate and falcate, with the upper lobe long and the lower lobe slightly shorter. The first dorsal fin is large, with a convex anterior margin, a slight concave posterior margin, and a pointed apex. However, the apex of the first dorsal fin can be rounded in newborn white sharks. The origin of the first dorsal fin is over the inner margin of the pectoral fins. The second dorsal fin is very small. The anal fin is approximately the same size as the second dorsal fin, with the origin of this fin slightly posterior to the second dorsal fin origin. The pelvic fins are small, but larger than the anal fin. The pectoral fins are shaped like scythes, with their anterior margin convex with the posterior margin moderately concave and the apex pointed.

The caudal peduncle, or base of the tail, is expanded laterally forming strong dermal keels. Precaudal pits are present on the upper and lower sides of the caudal peduncle, close to the origin of the caudal fin (Bigelow and Schroeder, 1948; Compagno, 1984; De Maddalena, 2002; Last and Stevens, 1994).

Similar Species

The identity of the great white shark is unmistakable. Nevertheless, there are some other sharks that show strong similarities with this species because they are closely related. These species belong to the genera *Isurus* or *Lamna*, and belong to the family Lamnidae, which includes the shortfin mako *Isurus oxyrinchus*, longfin mako *Isurus paucus*, porbeagle *Lamna nasus* and the salmon shark *Lamna ditropis*. All of these species have been recorded in U.S. waters.

These sharks attain a large size and have a spindle-shaped body, conical snout, dark circular eyes, prominent lower teeth, caudal keels, and a lunate caudal fin. The similarities among the species of the family Lamnidae are not so strong as to create prob-

1— Biology, Ethology and Ecology of the Great White Shark

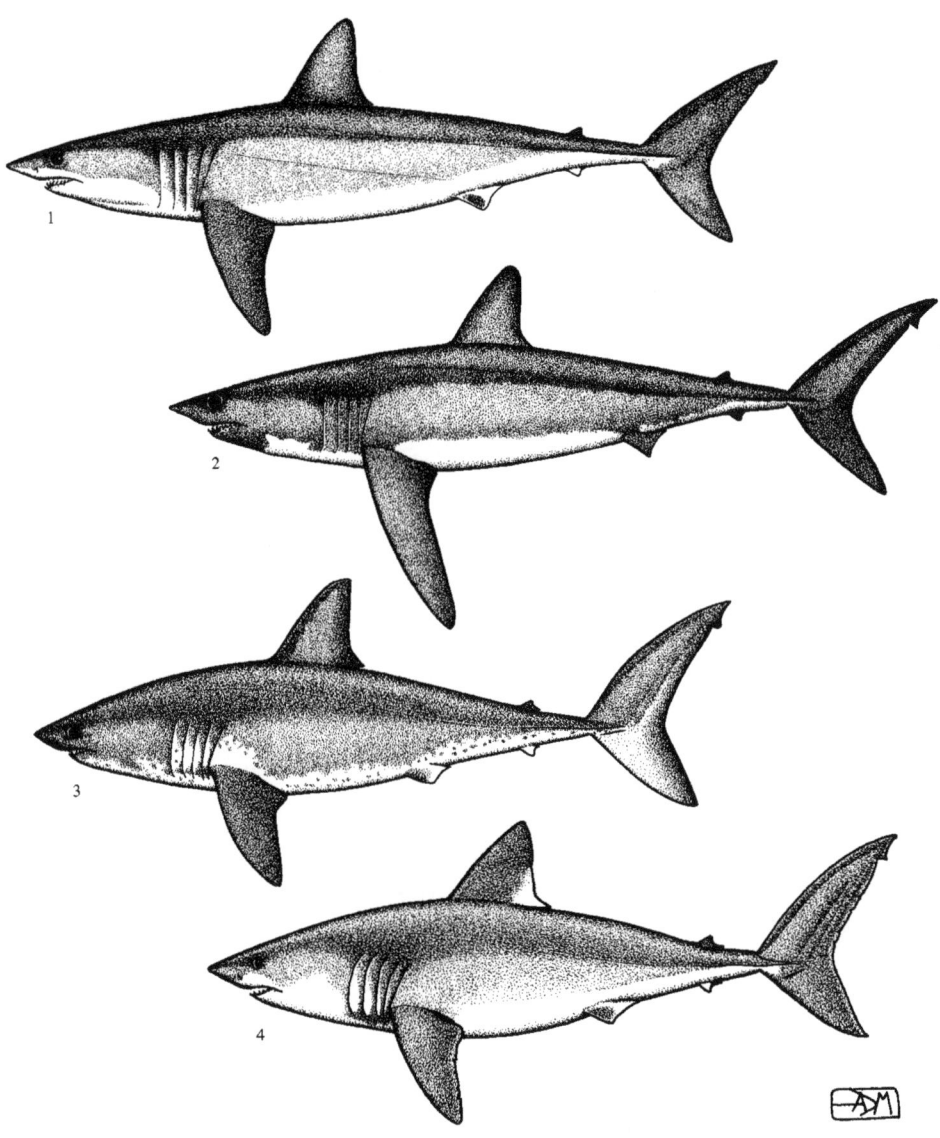

There are some other sharks that show strong similarities with the great white shark, because they are closely related. The family Lamnidae includes four other species: shortfin mako shark *Isurus oxyrinchus* (1), longfin mako shark *Isurus paucus* (2), salmon shark *Lamna ditropis* (3) and porbeagle *Lamna nasus* (4). All these species have been recorded in U.S. waters.

lems in identification, except when a novice observer examines a very young white shark. The sharks of the genus *Isurus* (*Isurus oxyrinchus* and *Isurus paucus*) have a slender body and a narrow snout. The teeth are strongly pointed, curved and narrow with cutting margins without serrations. The coloration can be bright blue with strong metallic reflections (*Isurus oxyrinchus*) or grey-blue to black (*Isurus paucus*). The other

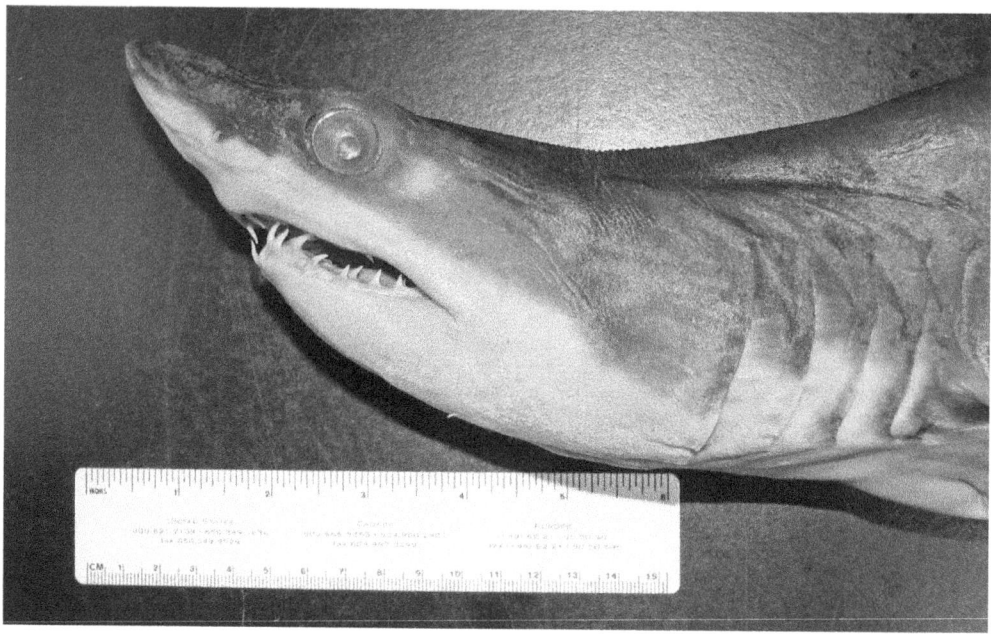

A shortfin mako *Isurus oxyrinchus* preserved at the University of Washington Fish Collection in Seattle, Washington, with cat. no. UW 047593 (photograph by Jeff Benca, courtesy University of Washington Fish Collection, Seattle, Washington).

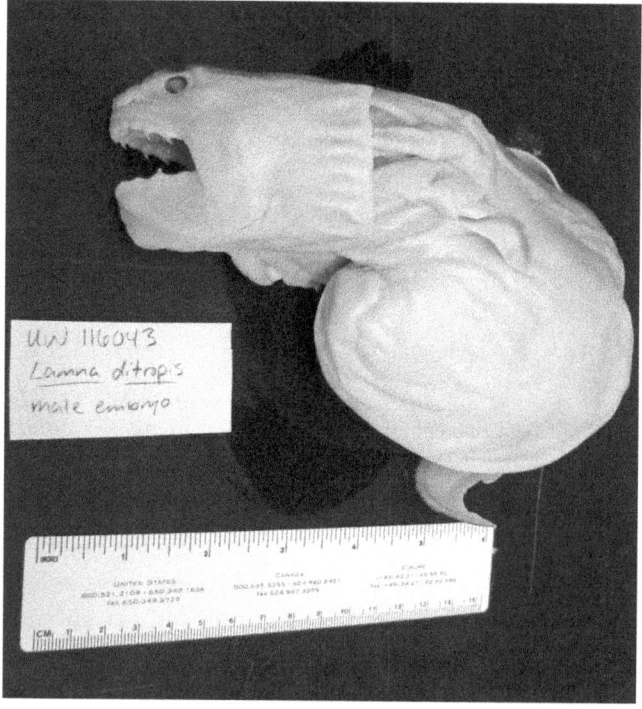

sharks of the genus *Lamna* (*Lamna nasus* and *Lamna ditropis*) have two pairs of caudal keels that are also present in *Isurus paucus* but not in *Isurus oxyrinchus*, which has a single pair. The teeth are small, lack serrations and have two small lateral cusplets. The coloration can be bluish-grey to black and, in *Lamna nasus*, a free

An embryo salmon shark *Lamna ditropis*, preserved at the University of Washington Fish Collection in Seattle, Washington, with cat. no. UW 116043 (photograph by Jeff Benca, courtesy University of Washington Fish Collection, Seattle, Washington).

1— Biology, Ethology and Ecology of the Great White Shark

rear tip of the first dorsal fin has a conspicuous white patch (De Maddalena, 2007; De Maddalena, *et al.*, 2005; De Maddalena and Baensch, 2008).

Hydrodynamic Design and Swimming

The great white shark planes gracefully through the water on its broad pectoral fins by slow undulations of its tail, while keeping the rest of its body rigid. The caudal fin is used for propulsion as the great white shark swims by moving its caudal fin from side to side. The pectoral fins, and to a lesser degree the pelvic fins, provide support and stabilization. The dorsal fins (in particular the first dorsal), and the anal fin, have a stabilizing function. In most sharks the longer upper caudal fin lobe drives the shark down during swimming, but this is balanced by lift generated from the slightly flattened head and the almost horizontal wide pectoral fins. But in the great white shark, the upper lobe of the caudal fin is only slightly longer than the lower lobe, and therefore this phenomenon is much less pronounced than in other shark species that have a

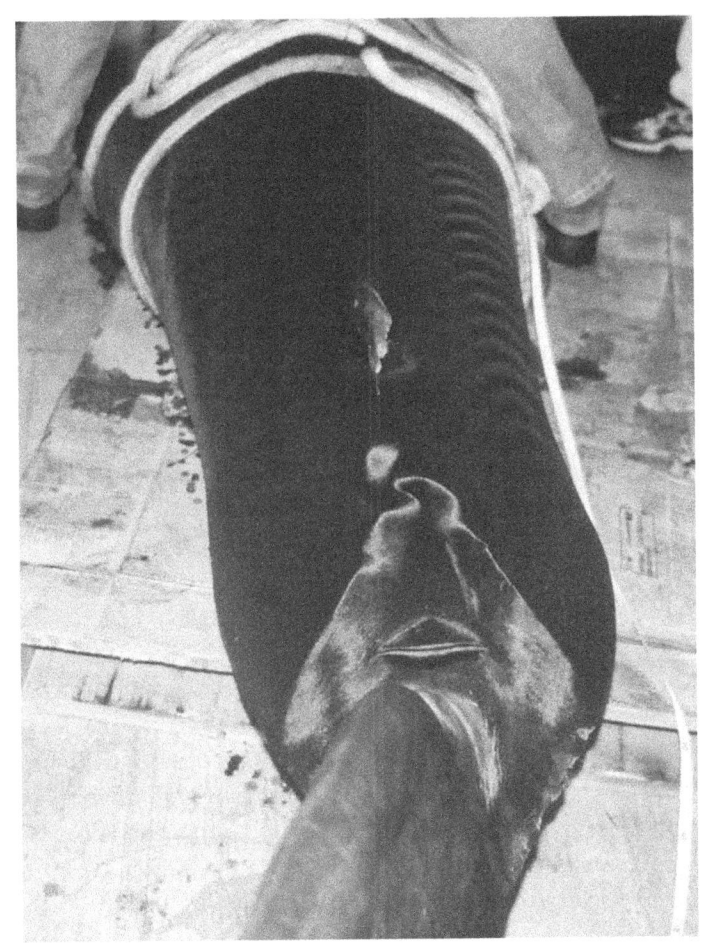

The caudal peduncle, caudal keels and precaudal pit of a 433 cm TL female great white shark that was captured on November 23, 2000, off Morro Bay, California. Sean R. Van Sommeran of the Pelagic Shark Research Foundation has also observed the presence of a "V" shaped area surrounding the precaudal pit, which is evident in this photo. Van Sommeran has hypothesized that this feature may be related to tail hydrodynamics (courtesy Pelagic Shark Research Foundation).

strongly asymmetrical caudal fin. The head of a great white shark is not dorso-ventrally flattened like in most sharks but is conical in shape.

The great white shark has a highly hydrodynamic design based on morphological features, such as a streamlined and spindle-shaped body, a conical snout, lateral keels on the sides of the caudal peduncle and precaudal pits on the upper and lower sides of the caudal peduncle (see page 12, "External Anatomy"). The caudal peduncle, or base of the tail, is expanded laterally, forming strong dermal keels which stiffen and streamline it laterally. Precaudal pits are grooves of uncertain function, which may be related to dorsoventral flexing of the caudal fin or to lateral tail hydrodynamics (Compagno, 1999).

Despite the excellent hydrodynamic design for straight line swimming and its rather stout body that may appear quite rigid, the powerful flexible tail and large pectoral fins give this shark excellent maneuverability. The great white shark can quickly turn horizontally and even pitch up or down to change depth quickly.

Because lamnid sharks exhibit regional endothermy (see page 28, "Circulatory System and Body Temperature"), the great white shark is an active and powerful swimmer. This predator can swim relatively slowly for long periods, with an average swimming speed of 3.2 km/h (Carey *et al.*, 1982). Nevertheless, it is capable of rapid acceleration, suddenly reaching high speed, which has been estimated in short bursts to reach approximately 40 km/h or more. It can also jump out of the water, reaching at least 3 m in height either during an attack, or as a form of play (De Maddalena, 2007, 2008; Martin, 2003). It is interesting to note that this behavior is frequently observed in the waters of False Bay, South Africa, but is unusual in other areas such as in the Mediterranean Sea (De Maddalena, 2009).

Measurement and Estimation of Size

Various measurements of size have been used in the past for the great white shark. The length has been reported both as straight line length and measured over the curvature of the body. However, the accepted scientific measurement is straight line length. The total length (TL) is the maximum length of the shark measured from the tip of the snout to the extremity of the upper lobe of the caudal fin. The total length can be measured with the caudal fin in the natural position (TLn) or with the caudal fin in the depressed position (TOT). The fork length (FOR or FL) is the distance from the tip of the snout to the fork between the lobes of the caudal fin. The precaudal length (PRC or PL) is the distance from the tip of the snout to the origin of the caudal fin.

Length-to-length and length-to-weight relationships for great white sharks have been reported in various studies and the equations are as follows:

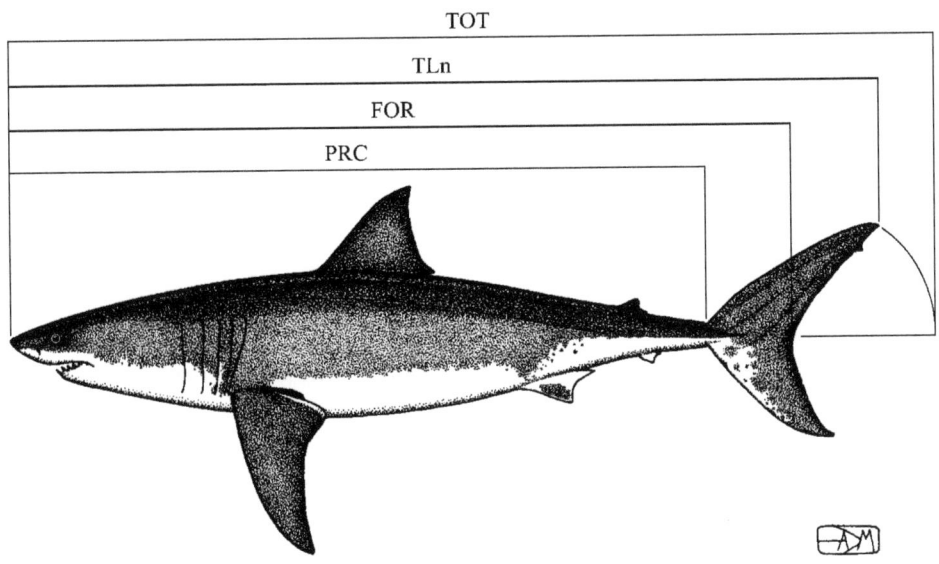

The total length of the shark is measured from the tip of the snout to the extremity of the upper lobe of the caudal fin. The total length can be measured with the caudal fin in the natural position (TLn) or with the caudal fin in the depressed position (TOT). The fork length (FOR or FL) is the distance from the tip of the snout to the fork between the lobes of the caudal fin. The precaudal length (PRC or PL) is the distance from the tip of the snout to the origin of the caudal fin.

$W(kg) = 3.8 \times 10^{-6} \times TL(cm)^{3.15}$ (n = 127) (Tricas and McCosker, 1984);

$W(kg) = 4.804 \times 10^{-6} \times TL(cm)^{3.095}$ (n = 200) (Casey and Pratt, 1985);

$W(kg) = 1.84 \times 10^{-5} \times PCL(cm)^{2.97}$ (n = 309, PCL = 131 to 348 cm) (Cliff *et al.*, 1989);

$W(kg) = 2.14 \times 10^{-5} \times PCL(cm)^{2.944}$ (n = 383, PCL = 131 to 373 cm) (Cliff *et al.*, 1996);

$W(kg) = 7.5763 \times 10^{-6} \times FL(cm)^{3.0848}$ (n = 125, FL = 112 to 493 cm) (Kohler *et al.*, 1996);

$FL(cm) = 0.9442 \times TL(cm) - 5.7441$ (n = 112, TL = 122 to 517 cm) (Kohler *et al.*, 1996);

$TL(cm) = 1.245\, PCL(cm) + 7.975$ (n = 40, PCL = 131 to 373 cm) (Cliff *et al.*, 1996);

$FL(cm) = 1.100\, PCL + 3.554$ (n = 142, PCL = 131 to 373 cm) (Cliff *et al.*, 1996);

$W(kg) = 3.026 \times 10^{-6} \times TL(cm)^{3.188}$ (n = 156) (Compagno, 2001);

$W(kg) = 7.914\, TL(m)^{3.0958}$ (n = 327 TL = 96 to 465 cm) (Mollet and Cailliet, 1996);

$W(kg) = 37.73(G[m]^2 TL[m])^{0.9334}$ (n = 140) (G = girth) (Mollet and Cailliet, 1996).

Methods for obtaining the length of great white sharks from commonly preserved skeletal parts (teeth, jaws, vertebrae) have been investigated by various authors (De Maddalena, 2000b, 2002; Gottfried *et al.*, 1996; Mollet *et al.*, 1996; Randall, 1973, 1987; Zuffa *et al.*, 2002). Mollet *et al.* (1996) and De Maddalena (2007) found that the size of the largest teeth is a reliable index for estimation of the size of a very young shark, but it cannot be used reliably to obtain the size of medium-sized or large individuals. The size estimated from the upper jaw perimeter has been considered inaccurate (De Maddalena, 2000b, 2002, 2006b; 2007, 2009; Mollet *et al.*, 1996). Consequently, these methods have not been used in this work. The size of a great white shark can be estimated from the diameter of the vertebral centrum. However, there is insufficient evidence that this method is reliable. Considering the failure of the attempts to estimate the white shark's size from teeth and jaws, it is doubtful that accurate correlations for length can be obtained from the utilization of other skeletal parts. Given the rarity of cases in which the largest vertebral centra of a great white shark of unknown size was preserved, the correlation between white shark size and vertebral centra size is not useful.

Photographic documentation has sometimes been used to estimate the size of large white sharks, and then to validate or refute the size reported by the original source or eyewitnesses. The literature on great white sharks is rich with gross estimates obtained from photos not suitable for this kind of evaluation (De Maddalena, 2009). As it has

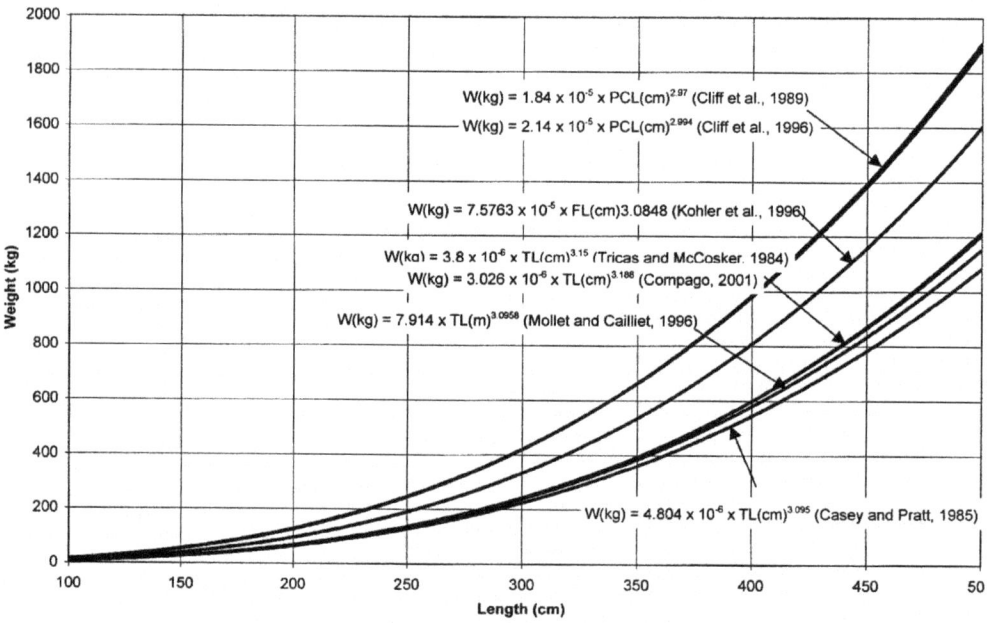

Length-to-weight relationships for great white sharks.

been pointed out in De Maddalena *et al.* (2001), the photos have to be used with great caution to estimate great white shark size, since the probability of obtaining inaccurate results is high. It is absolutely necessary to use only those rare pictures where the whole specimen or a wide part of it is not deformed by the lens utilized by the photographer or by the perspective. It is also essential that a large object of known size is located in the same plane of the shark in the photograph. All photos that do not meet this criteria have to be considered of no value for producing a valid estimate of the specimen size.

A 589 cm TOT female great white shark captured off Maguelone and landed in Sète, France, on October 13, 1956, is the largest specimen for which complete morphometric measurements are available (De Maddalena *et al.*, 2003). This morphometric data can be used to produce reliable estimates of large white sharks of equivalent size, approximately 600 cm in length, if suitable photographic documentation exists (De Maddalena *et al.*, 2001; 2003).

Maximum Size

As with most shark species, female great white sharks attain larger sizes than males. The maximum size of the great white shark has long been debated and used to be a subject of controversy. Today we have solid evidence that great white sharks attain 668 cm TOT, and probably even over 800 cm TOT (De Maddalena *et al.*, 2001).

The 589 cm TOT female great white shark captured off Maguelone and landed in Sète, France, is the largest preserved specimen available today and is located in the Museum of Zoology in Lausanne, Switzerland (without a catalogue number). This specimen is preserved as the largest cast in the world, and was reconstructed directly from a whole specimen. The shark was accurately measured at the time of capture and is still measurable today from the cast, making it the largest specimen ever reported with a size that cannot be disputed (De Maddalena *et al.*, 2003). A recent analysis of sighting and capture data for great white sharks from the Mediterranean Sea was presented in De Maddalena (2009). Of the 549 records of great white sharks, 74 specimens were reported to be measured or estimated as being larger than the 589 cm TOT, the size of the Maguelone specimen. In most cases the reliability of the reported size is impossible to verify, and cannot be accepted or refuted. Nevertheless, the size reported for some of these specimens needs to be considered reliable, even on the basis of the available photographic evidence.

The most reliable records of the largest great white shark measuring over 600 cm in total length are those that follow, in order of decreasing size. A specimen caught off Dakar, Senegal, in 1982, was estimated to be longer than 800 cm TLn, and even if it was never accurately measured, the size has to be considered perfectly reliable, having

This 589 cm TOT female great white shark captured off Maguelone and landed in Sète, France, on October 13, 1956, is the largest preserved specimen available today, and is located in the Museum of Zoology in Lausanne, Switzerland (without a catalogue number). This specimen is preserved as the largest cast in the world, which was reconstructed directly from a whole specimen. It also features the original fins and teeth (photograph by Michel Krafft, courtesy Musée cantonal de Zoologie, Lausanne, Switzerland).

been estimated by Juan Antonio Moreno, one of the major European shark experts (Barrull and Mate, 2001; De Maddalena *et al.*, 2001). A female caught near Filfla, Malta, on April 17, 1987, was estimated to be 668–681 cm TOT (De Maddalena *et al.*, 2001). Off Ganzirri, Italy, on June 19, 1961, a captured female was estimated at 666 cm TOT (De Maddalena *et al.*, 2001). A female caught near Kangaroo Island, Australia, on April

1— Biology, Ethology and Ecology of the Great White Shark

This female great white shark caught off Ganzirri, Italy, on June 19, 1961, was estimated to be 666 cm TOT (photograph courtesy Domenico Sorrenti).

1, 1987, was estimated 645 cm TOT (De Maddalena *et al.*, 2001). A female caught off Castillo de Cojimar, Cuba, in 1945, was reported measuring 640.8 cm, but again it was not known whether the length was measured in a straight line as TOT or as TLn (Bigelow and Schroeder, 1948; Guitart-Manday and Milera, 1974). A female caught off Mallorca, Spain, in March 1969, was estimated at 620 cm TL (Morey *et al.*, 2003) and a female caught off Mallorca, Spain, on February 5, 1976, was estimated to be 610 cm TL (Morey *et al.*, 2003).

Color

Like most sharks and other marine animals, great white sharks have dark backs, lighter flanks and white lower surfaces. This color pattern, called countershading, serves to render the shark almost invisible to their prey at a distance when viewed from above, from the side and from below. Moreover, the white lower surfaces are not visible from the side or above, so the contrast between the upper and lower coloration tends to be obscured. This is especially important for a predator that bases its predatory tactic on the element of surprise (see page 59, "Predatory Tactics") (De Maddalena, 2008). Besides camouflaging the shark, coloration plays an important role in species recognition.

A 274 cm TL female great white shark captured in June 2004 off the southern coast of California, was preserved at the Mote Marine Laboratory in Sarasota, Florida, with cat. no. GWS26. The grey area at the pectoral fin insertion can be clearly seen in this photograph (courtesy Mote Marine Laboratory and Aquarium, photograph by Charles Stevenson).

The dorsal coloration is intense deep blue, lead grey, brownish gray, or almost black along the back, and is a little bit lighter along the sides. The coloration is abruptly snow white on the undersides, with no color pattern. An irregular boundary separates the dorsal dark coloration from the ventral white coloration. A black or grey area is usually present at the pectoral fin insertion, also known as the pectoral fin axil. A very distinct black mark with an irregular margin is present on the ventral surface of the pectoral fins, at the apex. An irregular whitish area is present on the anterior margin of the caudal fin lower lobe. The dark dorsal coloration partially extends to the pelvic region by forming irregular patches. Newborn white sharks have a coloration that is very similar to that of the adults. So far, no difference in coloration has been described in different geographical areas. The coloration of the back and sides darkens to a sooty gray soon after death. A single case of albinism in *Carcharodon carcharias*, a specimen completely white with red eyes, has been recorded in South African waters (Smale and Heemstra, 1997).

Buoyancy

Most bony fish have a gas bladder, which is a gas-filled sac located in the upper part of the body cavity and is used to offset the weight of denser tissue such as bone.

The gas bladder is filled to the appropriate volume to render the fish neutrally buoyant. Sharks lack a gas bladder, but because of the light cartilaginous skeleton and a huge oily liver with a very low specific gravity, they are only slightly denser than sea water. Specific gravity is the ratio of the weight of a given volume of a substance to that of an equal volume of water. If the specific gravity of a pelagic organism is 1.0, it is neutrally buoyant. In fact, shark liver oil is five to six times less dense than sea water, having a specific gravity less than 0.20 (De Maddalena, 2008). The low density liver of the great white shark reduces the specific gravity of the whole shark by offsetting the higher density parts, like the musculature and the cartilage skeleton. The lower the shark density, the less lift from swimming is required to maintain its position in the water. The difference in the density of various sharks is also related to their habitat. Pelagic species are less dense than benthic species (Stevens, 1987). In the great white shark, the liver can account for up to 25 to 28.12 percent of the total body weight (De Maddalena, 2009).

It was once believed that sharks were less energy efficient than bony fishes because they lack a gas bladder. Actually, sharks can move up and down in the water column more easily than bony fish with swim bladders since they maintain close to neutral buoyancy through incompressible oily livers that are not affected by variations on pressure due to depth (Stevens, 1987). The air bladder of the bony fish is compressible, changing volume with depth, and needs to be filled or drained to maintain neutral buoyancy. Like all sharks, the great white shark is negatively buoyant and needs to swim constantly to maintain constant depth. If it stops swimming, it will sink slowly to the bottom.

Skin

The 10 mm thick skin from a 5.00 m long great white shark can weigh 120 kg (De Maddalena, 2007). Shark skin is covered with small "dermal denticles" or "placoid scales." In great white sharks, the dermal denticles are microscopic in size, with the mean size being 0.25 mm. The skin is almost smooth to the touch, only slightly rough and abrasive. A denticle has an inner core composed of a pulp, which is covered by dentine, which is covered by an outer layer of an enamel-like vitrodentine. The denticle has a bony basal plate or root that is set into the skin and a cusp at the opposite end. Dermal denticles reduce friction drag during swimming and noise generated by the great white shark's movement. Dermal denticles also have a protective function, since they cover the shark skin like armor, giving it strength. Denticles do not increase in size as the shark grows; instead, new denticles are added. The shape of dermal denticles varies from species to species and from body part to body part. In great white sharks, the denticles are densely distributed, and overlap each other along the anterior and lateral margins. The denticles have three crests and the posterior margin is smoothly

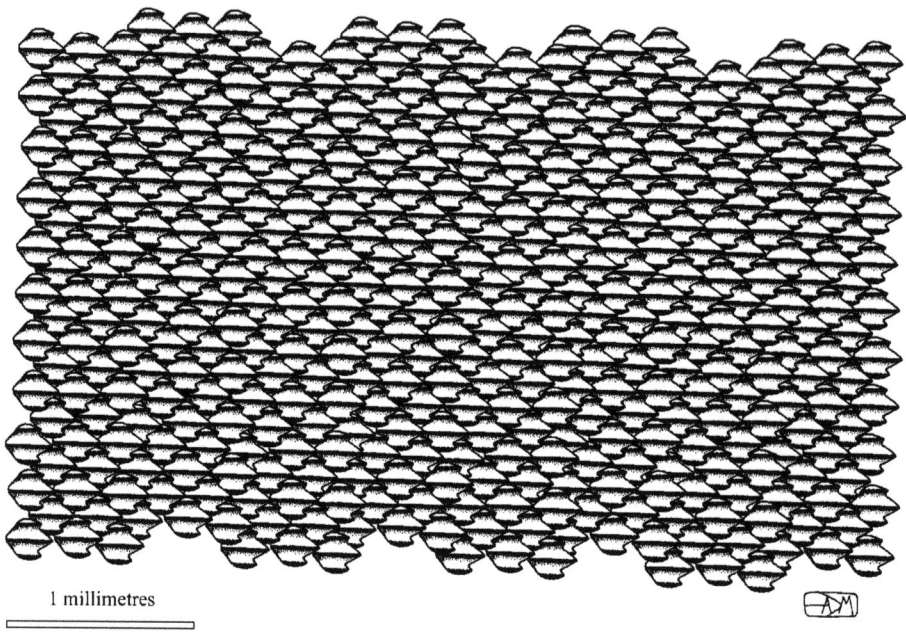

1 millimetres

Great white shark skin is covered with small "dermal denticles" or "placoid scales," with the mean size being 0.25 mm. The denticles are densely distributed, and overlap each other along the anterior and lateral margins. The denticles have three crests and the posterior margin is smoothly "W" in shape.

"W" shaped. The pedicle, or the stalk-like supporting structure of the cusp, is rather short and stout (Bigelow and Schroeder, 1948). The shape of dermal denticles is also important in the identification of a species. For example, identification of species by examination of dermal denticles has been used to identify which shark species are most impacted by the trade of their fins (Tanaka *et al.*, 2002) (see page 74, "Fisheries").

Respiratory System

Like other sharks, great white sharks employ gills for respiration. The mouth of the shark leads to the buccal cavity and to the pharynx where the gills are located. Oxygen is extracted from the sea water by highly vascularized membranes called gill lamellae, while carbon dioxide is expelled to the sea water. Like most other sharks, the great white shark has five pairs of gill slits. While bony fish gills are covered by a flap called an operculum, the gill slits of the shark are external (De Maddalena and Baensch, 2008).

Many sharks have a pair of spiracles, two small rudimentary gill openings, which are situated behind or below the eyes on each side. The spiracles are used instead of the mouth as an inlet for water, especially in sharks that live on the sea bottom. Therefore, they are larger in benthic species and smaller or absent in pelagic species. The

1 — Biology, Ethology and Ecology of the Great White Shark

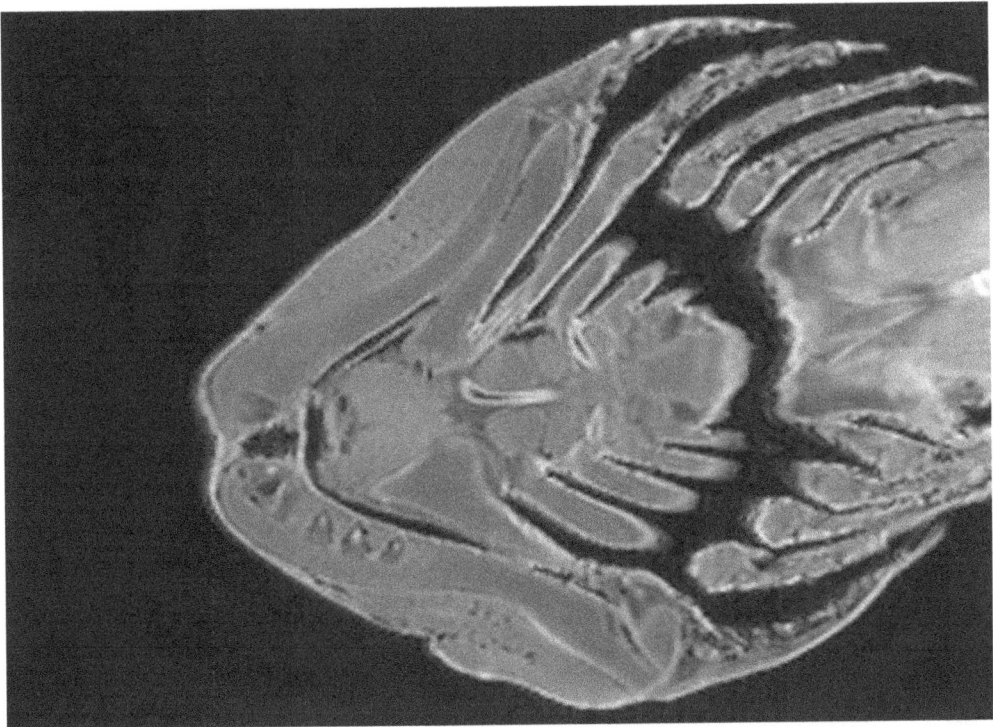

Horizontal magnetic resonance image slice of the head of a 130.0 cm TL great white shark captured on August 8, 2005, off Ventura, California, which was preserved as a whole specimen at the Scripps Institution of Oceanography in La Jolla, California, with cat. no. SIO 05-33. The visceral arches composing the splanchnocranium (mandibular arch, hyoid arch and gill arches) and the gill chambers are visible (courtesy Digital Fish Library/Center for Scientific Computation in Imaging/Center for Functional Magnetic Resonance Imaging/Scripps Institution of Oceanography/National Science Foundation).

spiracles are very small in great white sharks, being so minute that they can be seen only by close inspection (De Maddalena, 2007).

Forward motion may augment gill ventilation significantly. Great white sharks, and other sharks which swim continuously from birth until death, are considered to be obligate ram ventilators. This means that although some oral respiratory motion is present, they pass water over their gills primarily by opening their mouths while swimming. They lack the pumping apparatus necessary to move sufficient volumes of water over their gills while stationary (Hewitt, 1984). Great white sharks are among the most active species, requiring large amounts of oxygen, swimming constantly to stay alive. Many sharks can lie on the sea bottom for long periods. This behavior is common in the benthic species and has been observed in some pelagic species. The great white shark has never been observed lying on the sea bottom because its need for oxygen does not allow it to rest still on the bottom. This also explains why only 14.4 percent of great white sharks were still alive when found in the shark nets that protect swimmers from shark attacks along the coast of KwaZulu-Natal, South Africa (Cliff et al., 1996). For the same reason, this species is not suited for captivity, and to date no great white shark has survived in an aquarium for more than a few months. In fact, the longest period in captivity was recorded at the Monterey Bay Aquarium, California, for a young specimen that was hosted for 198 days, from September 2004 to March 2005 before being released in the wild (see page 181, "September 2004: Monterey Bay Aquarium, Monterey, California").

Circulatory System and Body Temperature

Like all sharks, great white sharks have a simple circulatory system. The heart lies in the pericardial cavity, which is ventral in the thoracic region and anterior to the pectoral fins. The heart is divided into two main chambers, the auricle or atrium and the ventricle, which are connected in a series. Downstream of the ventricle is the conus arteriosus and upstream of the auricle is the sinus venosus. Blood leaving the heart goes from the ventricle to the conus arteriosus and into the ventral aorta. It then moves into the branchial arteries and then to the capillaries located in the gills, where gaseous exchange occurs. The blood takes in oxygen and releases carbon dioxide. The blood is then collected in the dorsal aorta, and continues downstream through the rest of the body via smaller arteries. After the capillaries have delivered the oxygen and the nutrients to the organs and removed the waste products, including carbon dioxide, the blood enters the venous system. It then moves into the kidneys and releases waste products, with the

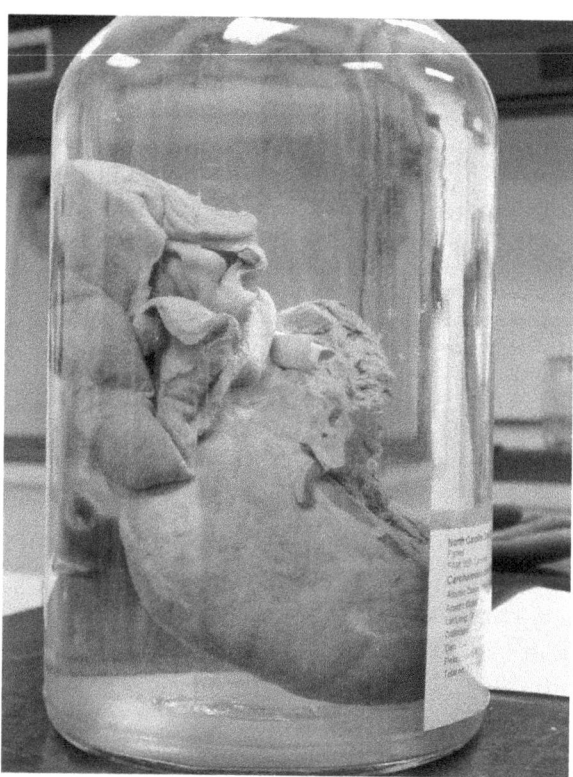

A heart from a 472.4 cm male great white shark captured on September 26, 1984, south of Beaufort Inlet, North Carolina, preserved at the North Carolina State Museum of Natural Sciences in Raleigh, North Carolina, with cat. no. NCSM 28360 (photograph by Gabriela Hogue, courtesy North Carolina State Museum of Natural Sciences).

exception of carbon dioxide. The blood then, primarily via the cardinal veins, returns to the heart by entering the sinus venosus and then the auricle (Castro, 1983; De Maddalena and Baensch, 2008; Randall, 1986).

While most sharks have body temperatures equal to that of the surrounding seawater, some species of the Order Lamniformes, in particular belonging to the family Alopiidae (thresher sharks) and to the family Lamnidae (great white sharks included), exhibit regional endothermy, enabling them to maintain a higher body temperature than that of the seawater because of a heat-retaining system. Red muscles are the most powerful during sustained swimming. Endothermic sharks have large amounts of red muscle tissue situated deep in the trunk, close to the vertebral column, while other species have these red muscles located closer to the skin. The red muscle tissue is connected to the circulatory system by a complicated network of arteries and veins called the "rete mirabile." The heat that is generated in the red muscles by swimming warms the blood. In the rete mirabile, the warm blood leaving the red muscle passes through the veins that carry blood from the organs to the heart. The heat is then transferred to the parallel arteries that carry cold blood from the gills to the organs. So heat is retained in the shark's body, rather than dissipating to the environment through the gills (Ellis and McCosker, 1991). This peculiar structure of the circulatory system allows the great white shark to maintain a body temperature between 4°C and 14°C higher than the ambient water temperature (Carey *et al.*, 1985; McCosker, 1987; Goldman *et al.*, 1996). Heat is a form of energy, so great white sharks have more energy at their disposal than most sharks that are cold-blooded. Chemical reactions are a function of temperature, being faster at higher temperatures. The muscles operate faster allowing the shark to swim faster than a fish that operates at a cooler temperature. Therefore great white sharks are very powerful, fast, capable of fast acceleration and able to leap high above the sea surface (De Maddalena, 2007, 2008).

Digestive System

As in other sharks, digestion in great white sharks takes place in the buccal cavity, pharynx, esophagus, stomach and intestine. The mouth of a great white shark is situated on the undersurface of the head, is very large, and has a parabolic curved shape. The upper and lower labial furrows at the corners of the mouth are very short, with the upper furrows shorter than the lower furrows.

The ventral position of the mouth is not an impediment to feeding as snout elevation and upper jaw protrusion (see page 37, "Skeletal System") project the mouth in an almost terminal position. The bite action of the great white shark is comprised of a sequence of jaw and snout movements. The snout lifts, followed by depression of the lower jaw and protrusion of the upper jaw. The lower jaw then elevates and the snout drops. The entire bite action lasts about 0.9 seconds. During multiple bites, the

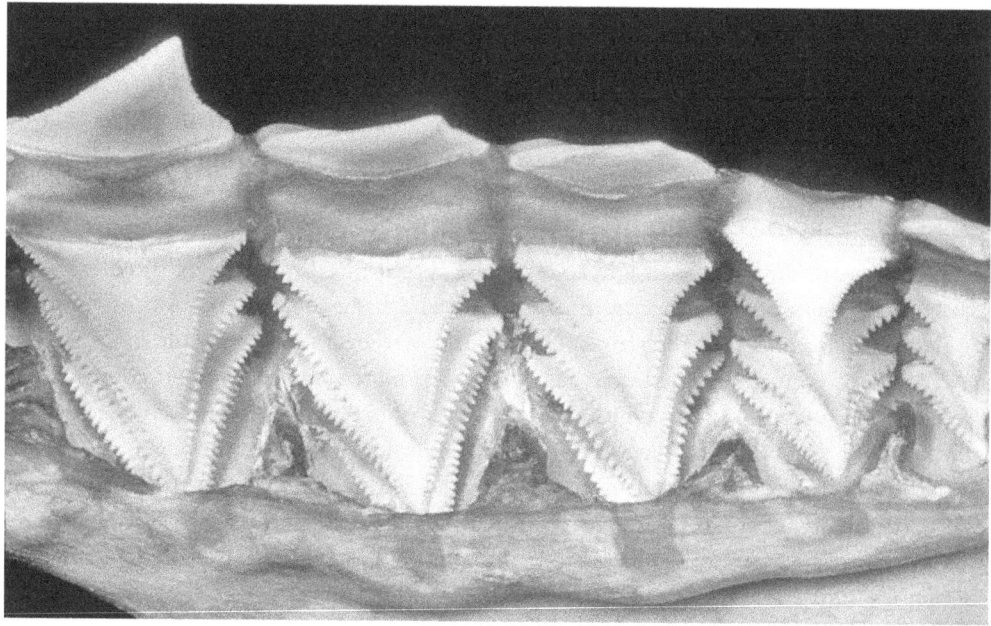

The upper-left jaw replacement teeth of a 249.2 cm TL female great white shark that was captured on July 12, 1959, off Goleta Point, California, and preserved at the California Academy of Sciences in San Francisco, California, with cat. no. CAS 26361. Each great white shark jaw typically has up to seven rows of teeth, but only the first and part of the second rows are functional (W.I. Follett, © California Academy of Sciences).

snout remains partially lifted (Tricas and McCosker, 1984). The great white shark removes large chunks of prey by biting it and at the same time shaking the head laterally. Large specimens can easily remove 20 kg of flesh in a single bite.

The teeth of sharks are merely modified and enlarged dermal denticles (see page 25, "Skin"), so it is not surprising that the teeth are almost identical in structure to the placoid scales. The teeth are composed, from the interior to the exterior, of a pulp, dentine and enamel-like vitrodentine over a bony base. Each tooth has a root and a crown, the projection of the crown being called the cusp. The crown has two margins, the lateral and the medial margin. The tooth is not fixed into a socket, but is implanted in the connective tissue (tooth bed) of the jaw by the root. Teeth are often broken and quite easily detached, but this is not a problem for the shark. The teeth are temporary as sharks have a perfect system of regular tooth replacement. Teeth are formed in a groove along the inner jaw surfaces, and behind the front teeth are several parallel rows of replacement teeth (De Maddalena and Baensch, 2008). Each great white shark jaw typically has up to seven rows of teeth (De Maddalena, 2000b), but only the first and part of the second rows are functional. As the front teeth are lost, replacement teeth rotate into position. Teeth are continuously replaced throughout life. In the young great white shark each tooth is replaced every 106.24 to 113.59 days, while in adults replace-

1 — Biology, Ethology and Ecology of the Great White Shark

ment slows, and each tooth is replaced every 225.90 to 242.18 days (Bruner, 1998). Considering that a great white shark has an average 48 teeth in its first row and that a maximum age estimate of 53 years has been calculated (see page 46, "Age"), great white sharks may shed over 5,000 teeth in a lifetime. This fact explains why fossil teeth of the genus *Carcharodon* are so common.

The number of front teeth in the upper and lower jaws is somewhat constant, and is used as an additional element of species identification. In order to represent the number of teeth in a shark mouth as a dental formula, the teeth in the first row of both jaws are counted. We refer to four quadrants when describing teeth: right upper jaw, left upper jaw, right lower jaw, and left lower jaw. The great white shark dental formula is usually 13 - 13/11 - 11, meaning that the great white shark usually has 13 teeth in each quadrant of its upper jaw, and 11 teeth in each quadrant of its lower jaw. To arrive at the total number of teeth in the outer row of upper and lower jaws, sum the numbers: 13 + 13 + 11 + 11 = 48. The dental formula often shows a pronounced variability. The great white shark formula has variability 12 to 14 - 12 to 14/10 to 13 - 10 to 13 (Cadenat and Blache, 1981). Recently, it has been pointed out that, because of its pronounced variability, the conventional dental formula is scarcely useful to identify shark species, and therefore more detailed descriptions are required, at least for the requiem sharks (Litvinov and Laptikhovsky, 2005). For the great white shark, the morphology of the

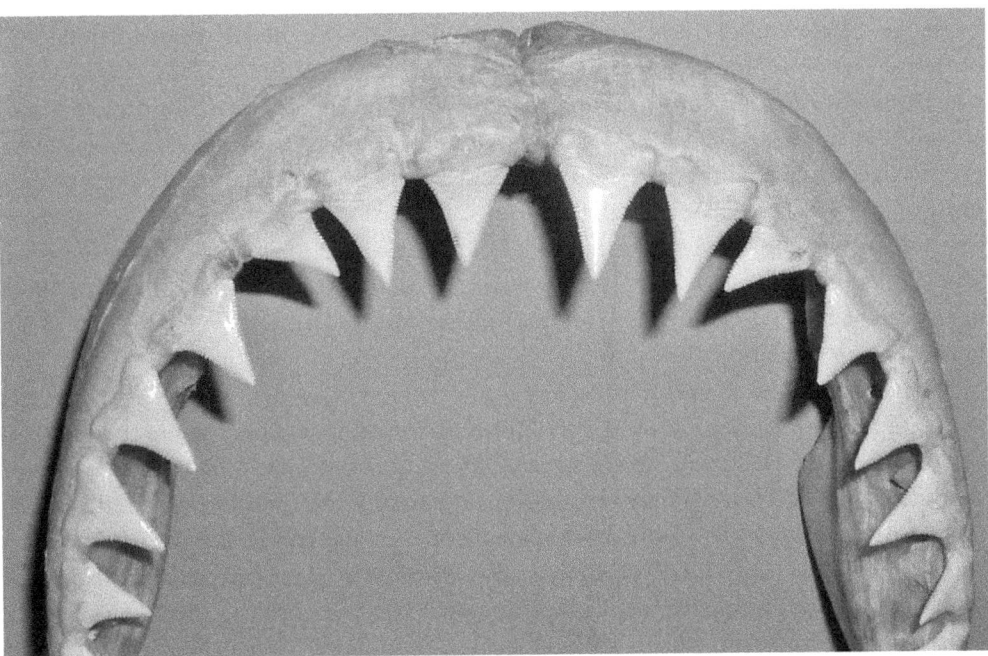

The teeth of the great white shark are large, triangular in shape, sharp with serrated edges, and are adapted for sawing pieces from large animals (photograph courtesy Antonella Preti, Natural History Museum of Los Angeles County).

teeth is so distinctive that it is very easily identified. The space at the symphysis is smaller in the upper jaw than in the lower jaw.

There are three main tooth shapes common to sharks with similar feeding strategies. The first group of teeth is adapted for sawing or shearing pieces from large animals. These teeth are large, triangular in shape, sharp, and with or without serrate edges, like those of the great white shark. The second group of teeth are adapted for seizing smaller, fast prey. These teeth are narrow and curved, moderately to very long, like those of the shortfin mako (*Isurus oxyrinchus*). The third group of teeth is adapted for crushing hard prey, such as mollusks and crustaceans. These teeth are smooth and arranged in a pavement formation, like those of the smooth-hounds (*Mustelus* spp.). The different species of sharks show many variations to these three main tooth shapes. In fact, teeth are an invaluable means of species identification (De Maddalena and Baensch, 2008).

Like most shark species, the great white shark's upper jaw teeth are different in shape from those of the lower jaw. The teeth of a great white shark have a large single cusp, triangular in shape and with strongly serrated margins. Great white shark teeth do not have cusplets, except in very young specimens. The teeth of newborn white sharks have two minute, pointed cusplets on each side of the cusp, and can also have a less marked serration or totally lack the serration (Uchida *et al.*, 1996). Lower teeth are slightly smaller and narrower than upper teeth. Teeth are classified based on position in the jaw. In the upper jaw, teeth are grouped as anterior, intermediate, lateral and posterior positions. In the lower jaw, teeth are grouped as anterior, lateral and posterior (Applegate and Espinosa-Arrubarrena, 1996). Anterior teeth are longer than others, both in upper and lower jaws. The subsequent teeth are successively smaller, with the smallest teeth near the corner of the mouth. The intermediate tooth, present only in the upper jaw, is smaller than the anterior tooth that is before it and the lateral tooth that follows it. The lower teeth stab and hold the prey securely while the upper teeth sever the flesh. The upper teeth of the great white shark are visible only when the shark opens its mouth. The lower teeth are prominent and well visible even when the shark has its mouth closed.

The great white shark's tooth shape is also related to its age. Tooth shape changes as the great white shark grows larger and feeds on different animals. As the shark grows, the teeth become thick and strong to accommodate larger prey. The male great white shark does not use its teeth only for feeding. Apparently, the male bites the female during courtship in order to stimulate the female to copulate and to keep her still during copulation. These "love bites" or mating scars are often evident on the flanks, belly, gill slits, back and fins (see page 44, "Reproduction") (De Maddalena, 2008). In addition to feeding and courtship, teeth are also used as weapons in defense against predators and for fighting individuals of the same species (see page 68, "Predators and Parasites," and page 66, "Competition").

The mouth, pharynx and esophagus are sufficiently wide to enable the animals to

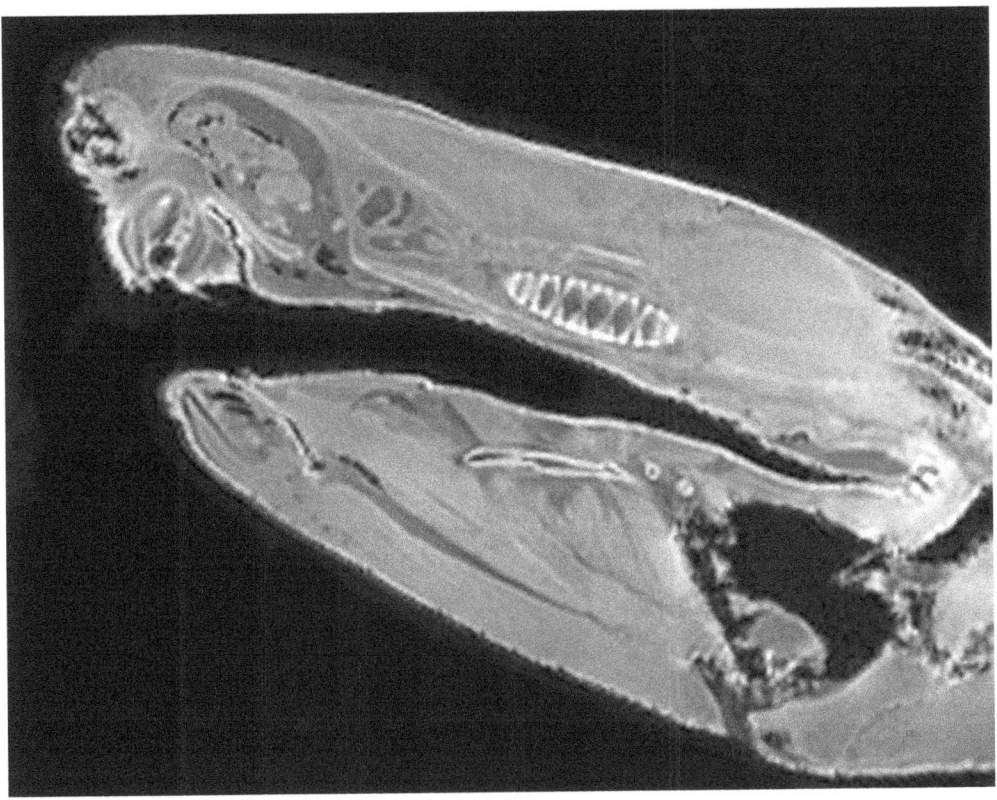

Sagittal magnetic resonance image slice of the head of a 130.0 cm TL great white shark captured on August 8, 2005, off Ventura, California, that was preserved as a whole specimen at the Scripps Institution of Oceanography in La Jolla, California, with cat. no. SIO 05-33. The buccal cavity and the pharynx are well visible (courtesy Digital Fish Library/Center for Scientific Computation in Imaging/Center for Functional Magnetic Resonance Imaging/Scripps Institution of Oceanography/National Science Foundation).

swallow large food items. Since the mouth leads to both the pharynx and gills, feeding and respiration in sharks are closely linked. The esophagus contains numerous internal fingerlike projections called papillae. Following the esophagus, the digestive system continues to the stomach. In the stomach, the food is broken down by enzymes. The stomach has longitudinal folds or rugae. Great white shark stomachs are "U" shaped, consisting of two portions: the cardiac stomach and the pyloric stomach. The upstream cardiac stomach is very large and saclike, while the downstream pyloric stomach is narrower. The shark stomach is large, enabling these formidable predators to ingest whole animals, large chunks of prey or a large amount of smaller prey (De Maddalena, 2008). Ingesting large amounts of food at a time means that they do not need to feed often. Moreover, food can be stored in the cardiac stomach undigested for long periods of time. Often, when a great white shark is eviscerated, prey is found in a near perfect state of preservation, intact or marked by only a few superficial teeth marks (De Maddalena, 2008). Sharks are able to evert the stomach, possibly in order to provide a means to empty it of indigestible objects. After eversion, the stomachs are usually returned to the interior of their body cavity, and the sharks remain in good health. This behavior has often been observed when sharks are captured (De Maddalena, 2008).

After the stomach, the digestive system continues with the intestine. The shark intestine is relatively short and the anterior part is the duodenum. As in humans, the duodenum, pancreas and liver are closely linked. The pancreas is an important digestive gland, which secretes digestive enzymes through the pancreatic duct into the duodenum. The liver of a great white shark (see page 24, "Buoyancy," and page 50, "Rate of Food Consumption") consists of two lobes, and is a huge organ that may comprise as much as 25 to 28.12 percent of the total body weight (De Maddalena, 2009). Sharks have large livers which store high-energy, fatty acids for buoyancy and for use as a food reserve. Liver cells secrete bile that emulsifies lipids. The bile is stored in a sac called the gall bladder. From the liver, the bile enters the hepatic duct, and from the gall bladder, the bile enters the cystic duct. These two ducts unite to form the common bile duct that discharges into the duodenum. The duodenum leads to the most important portion of the intestine, the ileum, which contains the intestinal valve. The intestinal valve is an internal structure that serves to increase the absorptive surface of the intestine without increasing the volume. This compact intestine provides great white sharks with the necessary space for a very large liver and stomach. There are three basic types of intestinal valves in sharks: the spiral valve, which resembles an auger in shape; the ring valve, which resembles a series of tightly-packed lamellae (plates) with a hole in their center; and the scroll valve, which resembles a loose roll of paper in shape. The great white shark has the ring valve, which has 47 to 55 plates very close to each other (Compagno, 2001). The absorption of the end products of digestion occurs in this section. The ileum leads to the last section of the intestine, the rectum. The rectal gland, which is located in this section, removes excess salt from the blood and opens by a duct into the rectum. The digestive system terminates with the anus that opens into the cloaca. In this small chamber, located on the ventral side of the animal between the two pelvic fin bases, the urinary and genital tracts also open to the outside (see page 41, "Excretory System," and page 42, "Reproductive System") (Castro, 1983; De Maddalena, 2008; Randall, 1986).

Nervous System and Sensory Perception

As in humans, the shark nervous system can be divided into two parts: the central nervous system, which includes the brain and the spinal cord, and the peripheral nervous system, which includes the cranial nerves and the spinal nerves. The brain is contained within the chondrocranium (braincase), and the spinal cord is protected by the cartilage of the neural arch of each vertebra (Randall, 1986) (see page 37, "Skele-

1 — Biology, Ethology and Ecology of the Great White Shark

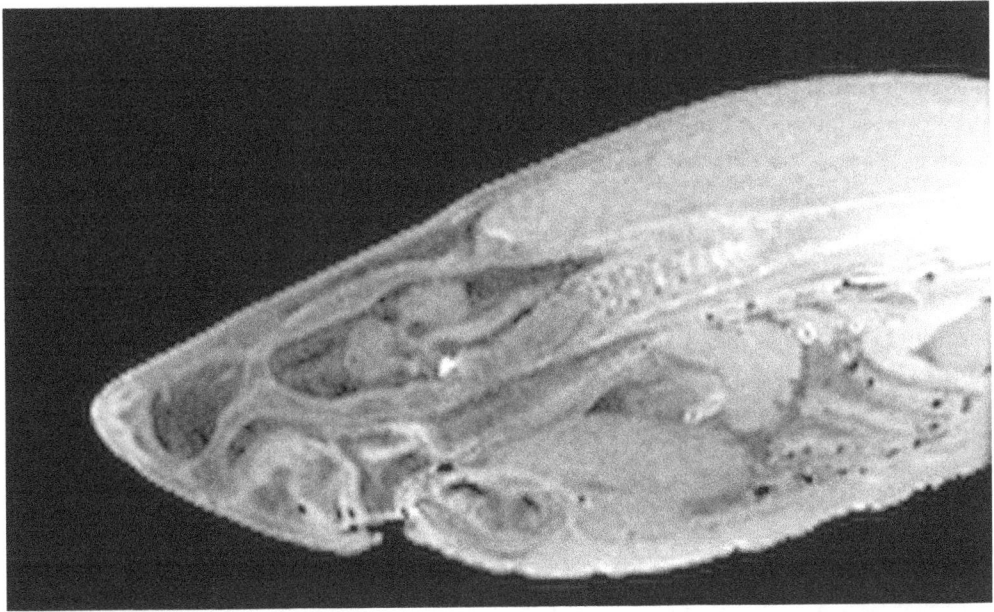

Sagittal magnetic resonance image slice of the head of a great white shark captured on June 13, 2006, off Long Beach, California, preserved at the Scripps Institution of Oceanography in La Jolla, California, with cat. no. SIO 06-78. The brain is partially visible inside the braincase or chondrocranium (courtesy Digital Fish Library/Center for Scientific Computation in Imaging/Center for Functional Magnetic Resonance Imaging/Scripps Institution of Oceanography/National Science Foundation).

tal System"). The complex brain of the great white shark must integrate a great deal of sensory information. Odors, sounds and similar low-frequency vibrations, minute electrical currents and visual stimuli can attract a great white shark. Sharks have four senses, which are chemoreception, mechanoreception, photoreception and electroreception. Great white sharks use each of these senses to locate prey, but different senses dominate at different times as they approach prey (De Maddalena, 2008).

At long distances, feeding areas are often located by olfaction. Sharks have a keen sense of smell. A bait slick, like a bleeding fish or a whale carcass, will soon attract great white sharks. The nostrils are partially covered by a nasal flap that separates sea water flowing in from water flowing out. The nostrils lead to an organ called an olfactory bulb that is composed of a series of tissue plates that are sensitive to chemicals. The olfactory bulb constantly receives a current of water while the shark is swimming (Hodgson, 1987).

The lateral line system and the ears enable sharks to detect movement in the sea water. The lateral line consists of a tube situated under the shark skin, along the flanks, extending from the snout to the caudal fin. The tube is connected with the exterior by short canals that connect to small pores on the skin surface. The tube contains sensory receptors called neuromasts, and each neuromast consists of a cluster of sensory hair cells. The lateral line is pressure-sensitive, enabling the shark to detect water vibrations. With this system, sharks are able to detect both the intensity of movement in the water and its direction from great distances. Low-frequency vibrations produced by wounded prey quickly attract great white sharks if they are within a hundred meters from the source. Great white sharks also use this lateral line to detect sea water currents (De Maddalena, 2008).

Great white sharks also have two inner ears, similar to those of humans, connected to the exterior on top of the head by narrow canals called endolymphatic ducts. The ear consists of three semi-circulars canals joined ventrally to three sacs, the utriculus, sacculus, and lagena, that regulate hearing as well as equilibrium (Randall, 1986). Hearing is similar to the lateral line system with hair cell receptors being situated in the inner ear. Hearing in sharks is very sensitive to low-frequency vibrations like those produced by wounded prey.

Prey detection and identification depend heavily on vision. The great white shark

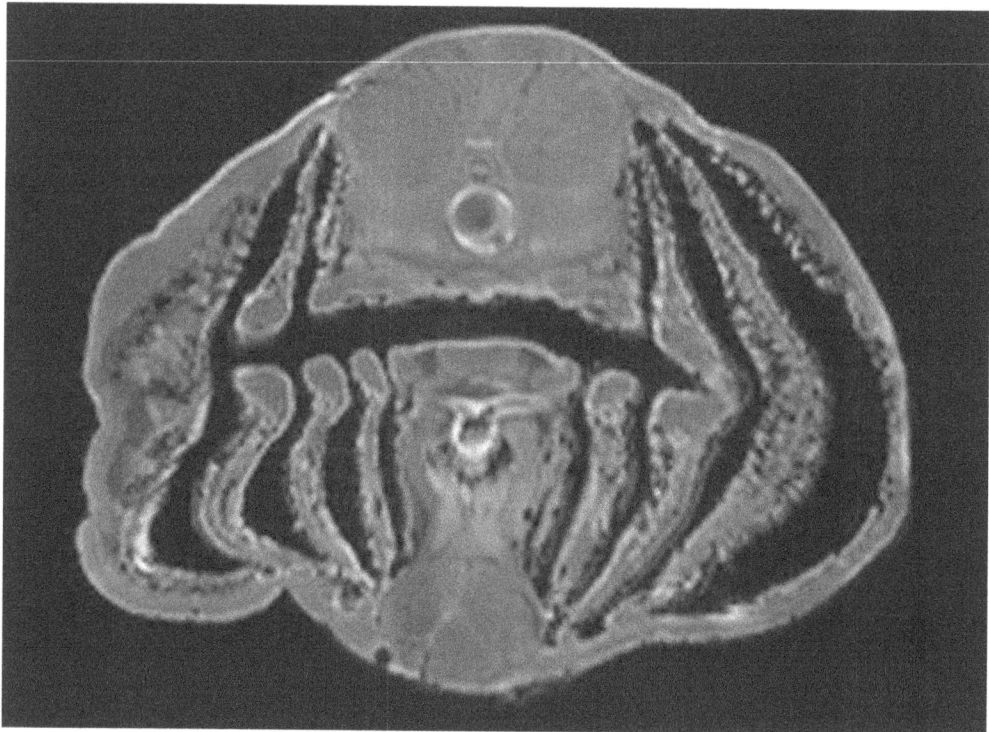

Axial magnetic resonance image slice of the head of a 130.0 cm TL great white shark captured on August 8, 2005, off Ventura, California, preserved as a whole specimen at the Scripps Institution of Oceanography in La Jolla, California, with cat. no. SIO 05-33. A vertebra with its neural arch containing the spinal cord is well visible (courtesy Digital Fish Library/Center for Scientific Computation in Imaging/Center for Functional Magnetic Resonance Imaging/Scripps Institution of Oceanography/National Science Foundation).

has relatively small eyes that are highly sensitive. Shark eyes are similar to those of many other vertebrates. The eyeball encloses the retina, which is the light-sensitive receptor area. Different parts of the retina are adapted for bright and dim light, allowing the eyes to function in low light conditions. The retina contains cone and rod photoreceptors. The cones function in bright light, while the rods function in dim light. The great white shark has acute color vision (Gruber and Cohen, 1978). The tapetum lucidum is a structure that lies under the retina and reflects incoming light back through the retina to restimulate photoreceptors, thus increasing the sensitivity of the eye (De Maddalena and Baensch, 2008). Most sharks have immoveable eyelids, but many sharks have a third eyelid, called the nictitating membrane, formed by an additional fold of the lower eyelid. The nictitating membrane is moveable, and when the shark is feeding the membrane closes over the eye to prevent damage (De Maddalena, 2008). However, the great white shark lacks the nictitating membrane, and in order to reduce the risk of injury, it rolls its eyes backward during an attack (Tricas and McCosker, 1984).

When great white sharks are within a few centimeters of their prey, they can detect the minute electrical currents generated by the nervous system of the prey by using electrical sensors called ampullae of Lorenzini. The ampullae are numerous small organs containing a sensory hair cell filled with an electrically conductive jelly. The external openings of the electroreceptors are small pores located over the head, and are particularly abundant on the underside of the snout (De Maddalena, 2008). These sensors are useful in the detection and taking of prey in the dark or when the prey is hidden under the sand of the sea bottom. Great white sharks also use this sense to orient themselves using the earth's magnetic field, particularly for long-distance migrations (see page 49, "Movements"). Great white sharks are also attracted to metals in response to the galvanic currents produced by electrochemical interactions between sea water and metals (De Maddalena, 2008).

During an attack, even touch and taste provide important sensory input. Touch receptors are located over the body of the shark, and this sense is used to obtain further information by means of bumping the prey. Gustatory receptors are located in the mouth and in the pharynx. Sometimes great white sharks decide on food palatability while it is lodged in their mouth (Collier *et al.*, 1996).

Skeletal System

Great white sharks, like all sharks, are cartilaginous fishes, having skeletons composed of cartilage. The only bony tissues are found in their teeth and scales. Cartilage is a light and flexible connective tissue that also forms the skeleton of the human embryo, but is predominately replaced by bone during development (cartilage persists in many areas of the adult human being, such as the nose, the ears, and the articular surfaces of

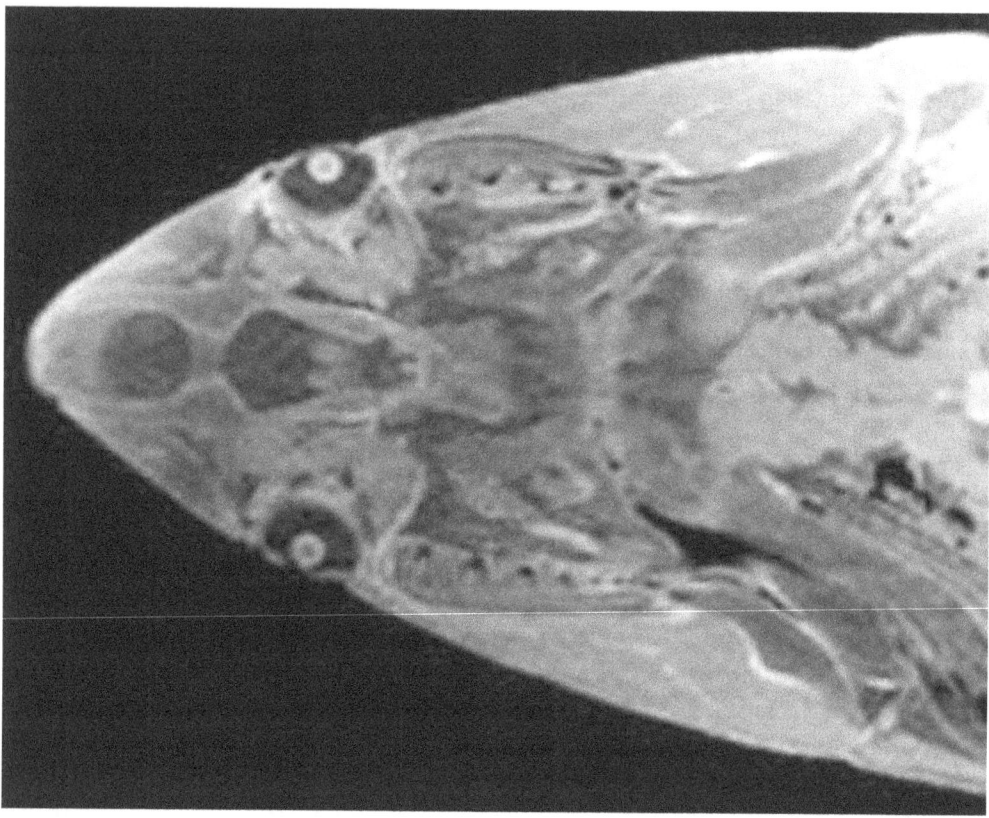

Horizontal magnetic resonance image slice of the head of a great white shark captured on June 13, 2006, off Long Beach, California, and preserved at the Scripps Institution of Oceanography in La Jolla, California, with cat. no. SIO 06-78. The whole skull is well visible (courtesy Digital Fish Library/Center for Scientific Computation in Imaging/Center for Functional Magnetic Resonance Imaging/Scripps Institution of Oceanography/National Science Foundation).

joints). Some of the cartilage in the shark body, such as the jaws, braincase and vertebrae, is partially calcified, hardened by the deposition of calcium salts. The cartilaginous skeleton of sharks is divided into axial and appendicular portions.

The axial skeleton consists of the skull and the vertebral column. The skull consists of a chondrocranium (also called neurocranium or braincase) and a splanchnocranium. The chondrocranium houses the brain (see page 34, "Nervous System and Sensory Perception"), and lacks the sutures and joints that separate the bones of the neurocranium in other vertebrates. The splanchnocranium is composed of the seven visceral arches: the mandibular arch (jaws), the hyoid arch and the five gill arches. Both upper and lower jaws are made of two cartilages that articulate at the symphysis, which is the joint along the median line at the front of the mouth. The upper jaw in sharks consists of the right and left palatoquadrate cartilages and the lower jaw consists of the right and left mandibular cartilages or Meckel's cartilages. The upper jaw does not have

1 — Biology, Ethology and Ecology of the Great White Shark

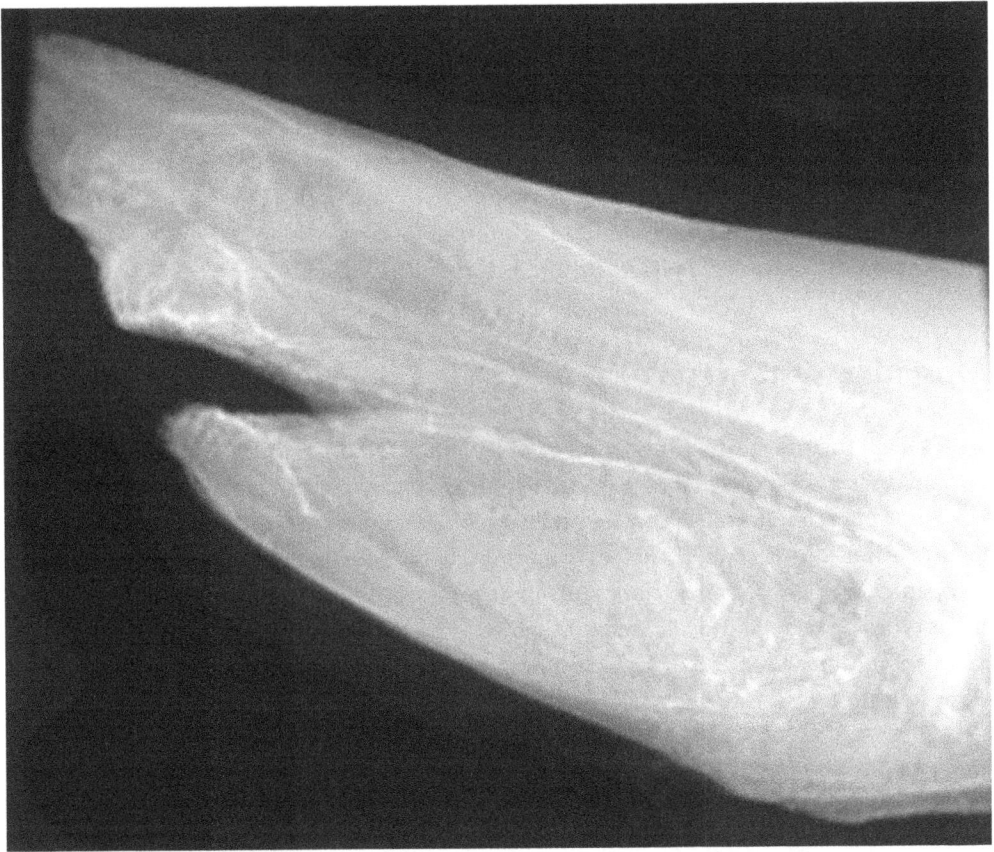

Sagittal magnetic resonance projection of the head of a 130.0 cm TL great white shark captured on August 8, 2005, off Ventura, California, and preserved as a whole specimen at the Scripps Institution of Oceanography in La Jolla, California, with cat. no. SIO 05-33. The vertebral column is well visible (courtesy Digital Fish Library/Center for Scientific Computation in Imaging/Center for Functional Magnetic Resonance Imaging/Scripps Institution of Oceanography/National Science Foundation).

a direct tight connection to the chondrocranium and is loosely suspended from it by ligaments. The upper cartilages of the hyoid arch are called the right and left hyomandibulars and form a bridge attaching the jaws to the chondrocranium. As a result, the upper jaw is highly mobile and protractible. This kind of jaw suspension is called hyostylic. It is thought that the jaws and hyoid arch have developed from primitive gill arches (Castro, 1983; Randall, 1986; Tortonese, 1956).

The vertebral column is made of cartilaginous vertebrae and extends from behind the head to the tip of the upper lobe of the caudal fin. The total number of vertebrae of the great white shark is 172–187 (Last and Stevens, 1994). A vertebra consists of three parts: the neural arch, the centrum, and the haemal arch. The neural arch is located at the top and encloses the spinal chord. The centrum is the central part, and encloses the notochord, which is the axial support of the body of the lower chordates and of the embryos of the higher chordates, the vertebrates. The haemal arch is located below, and protects major blood vessels (Randall, 1986).

The appendicular skeleton consists of the pectoral and pelvic girdles and the fin cartilages. These girdles do not articulate with the vertebral column, but "float" in the musculature. The fin cartilages provide support for the fins. They consist of cartilaginous supports known as basal and radial pterygiophores, which have parallel, elastoid rays known as ceratotrichia (Castro, 1983; Randall, 1986).

Muscular System

Like humans, great white sharks have skeletal, cardiac, and smooth muscles. Skeletal muscle is a type of striated muscle, which generally contracts voluntarily, although it can contract involuntarily through reflexes. Skeletal muscles connect to cartilage, typically with a joint between the cartilage members. Movement is created by applying force to cartilage via contraction, causing a rotation about the joint. Sharks have two main types of skeletal muscle, red and white. The red muscle has a good blood supply and uses aerobic oxidation of fat as its energy source. Red muscle functions in sustained slow swimming. White muscle has a poor blood supply, functions by the anaerobic

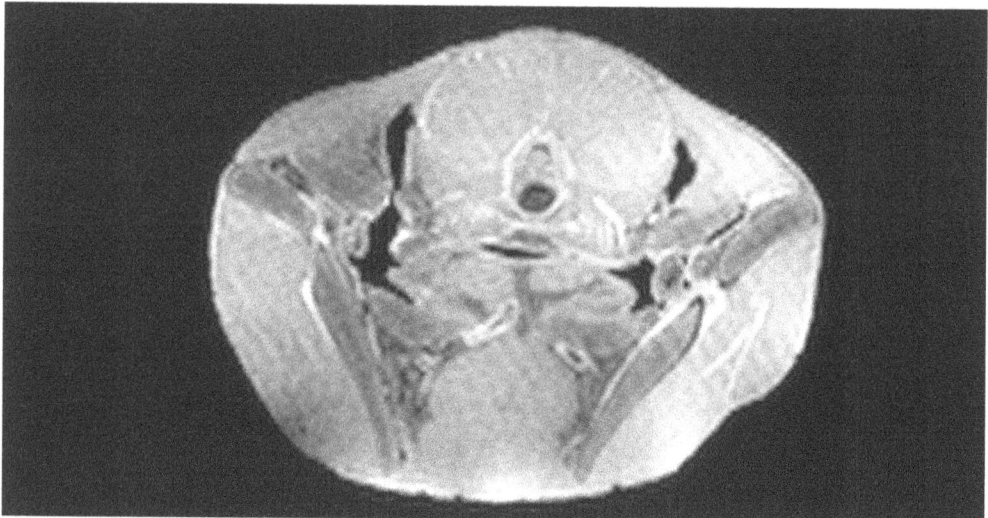

Axial magnetic resonance image slice of the head of a great white shark captured on June 13, 2006, off Long Beach, California, and preserved at the Scripps Institution of Oceanography in La Jolla, California, with cat. no. SIO 06-78. Note the voluminous musculature of the jaws, with the adductor mandibulae (courtesy Digital Fish Library/Center for Scientific Computation in Imaging/Center for Functional Magnetic Resonance Imaging/Scripps Institution of Oceanography/National Science Foundation).

breakdown of glycogen and is only used during fast sprint swimming. Because white muscle operates anaerobically, great white sharks cannot sustain sprint speeds and quickly become exhausted (Stevens, 1987). In most sharks, the red muscle lies in a thin layer just under the skin and outside the white muscle. Endothermic sharks have larger amounts of red muscle tissue located deep in the trunk, close to the vertebral column. The red muscle tissue is connected to the circulatory system by a complicated network of arteries and veins, the rete mirabile, which acts like a counter-current heat exchanger. Heat that is generated in the red muscles by swimming warms the blood. This outgoing warm blood transfers heat to the incoming cold blood as it flows out of the muscle, keeping the red muscle warm (see page 28, "Circulatory System and Body Temperature"). The main mass of skeletal muscular tissue of the shark, which produces the propulsive force in swimming, is segmentally arranged. Each segment of muscle, called a myotome, has a dorsoventral zigzag pattern when viewed from the side. The myotomes are separated from one another by a white partition of connective tissue called the myoseptum. Running longitudinally along the middle of the side of the body is another sheet of connective tissue, the lateral septum, which divides the musculature into the dorsal epaxial portion and the ventral hypaxial portion. The epaxial portion attaches to the posterior part of the chondrocranium. The hypaxial portion attaches to the pectoral girdle. The fin muscles are outgrowths of the myotomes. In the branchial region, the most conspicuous muscles are the superficial constrictors, which compress the gill chambers, forcing water out of the buccal cavity through the gill and closing the gill slits. In the mouth region, the most important muscle is the adductor mandibulae, which strongly closes the jaws (Randall, 1986).

Cardiac muscle is a type of involuntary striated muscle, which contracts involuntarily and is found in the walls of the heart. Cardiac muscle shares similarities with skeletal muscle with regard to its striated appearance and contraction. Contraction of cardiac muscle propels blood from the heart to the ventral aorta (see page 28, "Circulatory System and Body Temperature").

Smooth muscle is a non-striated muscle which contracts involuntarily. Smooth muscle is fundamentally different from skeletal muscle and cardiac muscle in terms of structure and mechanism of contraction. Smooth muscles are responsible for the contraction of hollow organs, such as blood vessels, the gastrointestinal tract, or the uterus.

Excretory System

Together, the excretory system and the reproductive system form the urogenital system. The kidneys of sharks are long, slender, dark red organs which lie on either side of the vertebral column. They are located outside the body cavity and separated from it by a membrane, the peritoneum (Castro, 1983). The functional excretory structures of the kidney are the renal corpuscles, which consist of two structures, the glomerulus and the Bowman's capsule. The glomerulus is a knot of capillaries that extracts fluid from the blood by filtration pressure to the encapsulating Bowman's capsule. The fluid passes through tubules from the Bowman's capsule where essential elements of the fluid are reabsorbed, leaving the rest to pass out as urine (Randall, 1986). In the male, the kidneys are drained directly into the cloaca by the ureters. In the female, the ureters unite posteriorly to form the urinary sinus, which is enclosed by a conical projection, the urogenital papilla, which opens by a pore into the cloaca. In the female the anterior part of the kidney is very reduced, and urine is produced only in the posterior part (Castro, 1983).

The urine of sharks contains little urea as sharks are unique in retaining a high percentage of urea in their blood and tissue fluids (2.0 percent to 2.5 percent compared to 0.01 percent to 0.03 percent for other vertebrates). They maintain this high urea concentration, along with a high chloride concentration, to overcome the osmotic problem experienced by marine organisms whose body fluids are hypotonic (less salty) to seawater. Osmosis is the passage of water across a semipermeable membrane from a dilute solution to a more concentrated solution. Without the high urea concentration, they would lose water over membranous surfaces, such as the gills, to the sea. In contrast, marine bony fishes are hypotonic and combat this osmotic problem by oral intake of seawater and active excretion of the salt by special cells in the gills (Randall, 1986). Sharks do not need to drink sea water because water naturally diffuses into the body through external cell membranes by osmosis.

Reproductive System

In mature males, a pair of large, elongated testes can be seen at the anterior end of the body cavity above the liver. In immature sharks, the testes are an inconspicuous mass. Spermatozoa are produced in the seminiferous tubules in each testicle and conveyed through the ductus efferens into a highly convoluted epidydimis located along the vertebral column on either side of the dorsal aorta. The epidydimis passes sperm to the ductus deferens. In great white sharks, the sperm is enclosed in protective packets called spermatophores, containing a very large number of spermatozoa (Pratt, 1996). The function of the spermatophore is to protect the sperm and to prevent loss of sperm by leakage into the water during copulation. The ductus deferens is a spermatophore-forming organ (Castro, 1983). In immature males, the ductus deferens is a straight tube on the ventral surface of the kidneys. In sexually mature males, the anterior part of the ductus deferens is highly coiled, while the posterior part is a straight tube that enlarges to form the seminal vesicles. The posterior end of the seminal vesicles forms the sperm sacs, and the two sperm sacs unite posteriorly to form the urogenital sinus, a cavity enclosed by a conical projection, the urogenital papilla, which protrudes into the cloaca. The male shark has two claspers, tube-like copulatory organs that originate from the

1 — Biology, Ethology and Ecology of the Great White Shark

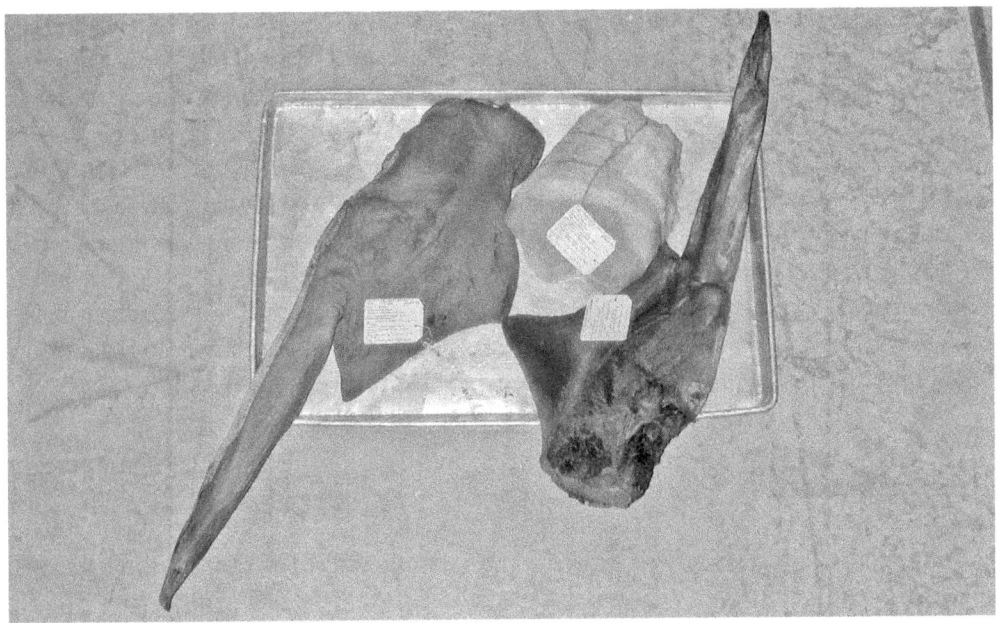

Claspers from a 472.4 cm male great white shark captured on September 26, 1984, south of Beaufort Inlet, North Carolina, and preserved at the North Carolina State Museum of Natural Sciences in Raleigh, North Carolina, with cat. no. NCSM 27589 (photograph by Gabriela Hogue, courtesy North Carolina State Museum of Natural Sciences).

medial margin of pelvic fins. Therefore, the sex is easily recognized by looking at the ventral side of the animal. In newborn male white sharks, the claspers are only a few centimeters long, while those in adult individuals are long and calcified. Males are considered to be sexually mature if sperm is present in the ampulla epididymis, if the claspers are calcified, and if the rhipidion, the head of the clasper, can be snapped open; in fact, in the mature males, the rhipidion can be opened, since there is a terminal spur that can be extruded (Pratt, 1996). During mating, one clasper is turned forward and is inserted into the female. Spermatophores are forced from the cloaca through the claspers and out into the female by means of a sea water current produced upon contraction of organs called siphon sacs (Castro, 1983).

In most female sharks, a pair of ovaries are located at the anterior end of the body cavity above the liver. But in the female great white shark, only the right ovary is well developed and functional, while the left ovary is vestigial (Ellis and McCosker, 1991). In immature females, the ovary is small and smooth, but in mature females it is large and with rounded protrusions from the developing ova. Each ovum is surrounded by a layer of nutritive cells called the follicle. When an ovum is ripe, it is discharged into the coelom by a rupture of the follicle and passes through the ostium, which is the opening of the oviduct. The ostium bifurcates into the right and left oviducts. The shell gland is an enlargement of the anterior part of each oviduct that secretes a membrane, enclosing the eggs as they pass through the oviduct. The shell gland also acts as the site for sperm storage and fertilization. The posterior part of each oviduct is enlarged to form the uterus, where the embryonic sharks develop. The two uteri unite posteriorly to form the vagina, which opens into the cloaca (Castro, 1983). Female sharks are considered mature if there is evidence of a current or previous pregnancy or evidence that they will be ready to reproduce within a short period of time. For females that are not or have not been pregnant, maturity can be determined by assessing the condition of the ova in the ovary and the size of the oviduct. Mature females will have well-developed yolky eggs in the ovary and the oviduct may start to enlarge and detach from the body wall. Females that have previously been pregnant will have an enlarged oviduct containing enlarged shell glands and well developed uteri (Conrath, 2005).

Reproduction

Sharks have a slow rate of growth, and consequently they also have long sexual maturation times. In great white sharks, males mature between 350 cm and 410 cm TL, which corresponds to between 9 and 10 years of age. Females mature from 400 cm to 500 cm TL, which corresponds to between 12 and 14 years of age (Compagno, 2001). Therefore, female great white sharks reach sexual maturity at a bigger size than males. Shark species that mature at a young age have a greater capacity to recover from exploitation than sharks that mature later (Smith *et al.*, 1998). Thus age at maturity is a crucial factor influencing the productivity of a species. The long sexual maturation time of the great white shark makes the species extremely vulnerable to overfishing.

White shark eggs are fertilized internally (see page 42, "Reproductive System") and mating should occur in the spring and summer months. White shark mating has been observed only once, in New Zealand waters. The male white shark began courtship by approaching the female and grabbing her with his mouth, causing her superficial wounds called love bites or mating scars. Then the male inserted one of his two claspers inside the female cloaca. The sharks stayed attached belly to belly, turning over from time to time. The copulation lasted some forty minutes (Francis, 1996).

Sharks exhibit one of the following three reproductive methods: (a) oviparous species lay horny egg cases containing embryos nourished by their yolk-sac; (b) aplacental viviparous species produce live young nourished in the uterus by a yolk-sac; (c) placental viviparous species produce live young nourished in the uterus by a placenta formed from a modified yolk-sac attached to the uterine wall.

Aplacental viviparity is the most common reproductive method in sharks. Great white sharks exhibit aplacental viviparity as the eggs are retained in the uterus throughout development. Fetus sharks obtain nourishment from their yolk sac and also by feeding on the unfertilized eggs produced by the mother (oophagy) (Saïdi *et al.*, 2005; Uchida *et al.*, 1996).

1 — Biology, Ethology and Ecology of the Great White Shark

The main great white shark nursery grounds for the Pacific U.S. waters are located inshore from Ventura Canyon in Southern California. A 130.0 cm TL very young great white shark captured on August 8, 2005, off Ventura, California, U.S.A., that was preserved as a whole specimen at the Scripps Institution of Oceanography in La Jolla, California, with cat. no. SIO 05-33. The item is not currently listed in the Scripps Institution of Oceanography collection database, with the exception of a small sample of muscle tissue with cat. no. SIO 05-66 (photograph courtesy Digital Fish Library/Center for Scientific Computation in Imaging/Center for Functional Magnetic Resonance Imaging/Scripps Institution of Oceanography/National Science Foundation).

Records of pregnant female white sharks are incredibly rare, which is a mystery. It is clear that the size of pregnant female white sharks make them hard to capture. It is possible that during the gestation period, the females abandon the rest of the population, going to remote areas far offshore, or into deep waters. It is also possible that they stop feeding temporarily when they give birth to their pups, so that the probability of contact with humans decreases (De Maddalena, 2002). These hypotheses are speculation and insufficient to explain this mystery. However, the rarity of pregnant female white sharks suggests that they have very low fecundity.

The gestation period is unknown but it may last over a year. The parturition may happen in spring or summer in temperate locations of both hemispheres (De Maddalena, 2009; Francis, 1996). Sharks produce relatively small numbers of young. Litter size of the great white shark ranges from 2 to 14, and perhaps up to 17 (Cliff *et al.*, 2000). At birth, pups emerge head first from the female cloaca. Like other sharks, great white sharks can give birth to their embryos in traumatic situations, such as when they are caught or are dying on a boat deck. After birth, the pups show an umbilical scar and are thus identified by the presence of an unhealed umbilical scar (Kabasakal, 2008). The size of great white shark pups ranges from 81.3 cm to 151 cm TL at birth (Ralph

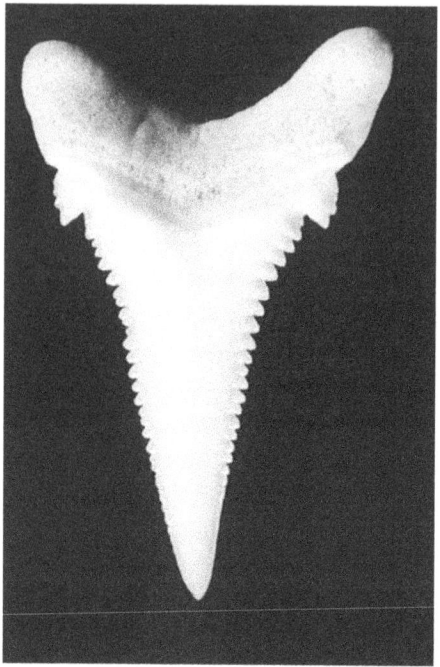

Tooth from a 167.6 cm TL female great white shark that was captured on September 9, 1936, off Malibu, California, preserved at the California Academy of Sciences in San Francisco, California with cat. no. CAS 25844. The teeth of very young white sharks have two minute pointed cusplets on each side of the cusp (W.I. Follett, Dept. of Ichthyology, California Academy of Sciences).

S. Collier, pers. comm.; Francis, 1996; Uchida et al., 1996). Great white shark pups receive no parental care, but are born fully formed and ready to feed. The size of white shark pups is quite substantial even at birth, which precludes them from being eaten by most predators. The mother does show particular attention to choose the birth area. Nursery areas, where pregnant females give birth and where juveniles remain for the early period of their lives, have been observed for the great white shark, similarly to many other species of sharks (De Maddalena, 2009). The abundance of suitable prey is the major reason some regions function as nursery areas for sharks (De Maddalena, 2008). Zones that have been found to be serving as great white shark nurseries, and probably breeding areas, include Southern California, Baja California and possibly Gulf of California, Mexico, the Mid-Atlantic Bight off the U.S. (New England and Mid-Atlantic States), the Channel of Sicily in the Mediterranean Sea, the east coast of South Africa from False Bay to KwaZulu-Natal, southern Australia, New Zealand, and Japan (Casey and Pratt, 1986; Compagno, 2001; De Maddalena, 2009; Cigala Fulgosi, 1990; Fergusson, 1996).

Age

Sharks live fairly long lives. In order to determine the age of a shark, several vertebrae are cut from its body and the central part is removed and treated to expose the rings. Various chemical processes can be used to accentuate growth patterns in the vertebrae of sharks. This can be done by etching the vertebrae with mild acid and silver nitrate, scorching and polishing. The patterns take the form of concentric bands, each made up of a number of closely spaced rings. The age of the shark can be estimated by counting these rings. However, there are usually many more rings than there are years that the shark has been alive, so there needs to be an understanding where one year of growth starts and finishes. The processes that govern vertebral growth have yet to be described in sharks. The pattern varies from one ring per year in most carcharhinids to two rings per year in some lamnids to the complete absence of periodicity (Skomal and Natanson, 2003). Captured sharks are now injected with chemicals like oxytetracycline that are quickly deposited in the vertebrae, creating a mark on the current ring. When sharks that have been treated in this way are recaptured, their vertebrae can be examined showing the band with the injected chemical, and any subsequent growth rings. Knowing the time between capture and recapture along with the number of rings deposited, the number of rings deposited per year can be determined. In great white sharks, a growth ring was defined as a band pair composed of one calcified (opaque) and one less-calcified (translucent) band. In this species, there is evidence of annual growth ring deposition, but definitive confirmation is still needed from further studies (Wintner and Cliff, 1999). The maximum age based on vertebral band pair counts is 53 years (Sabine Wintner, pers. comm. 2003).

Geographical Distribution

The great white shark is one of the widest-ranging chondrichthyians. It is found in the following areas. Western Atlantic: Newfoundland to Florida, Bahamas, Bermuda, Cuba, northern Gulf of Mexico; also Brazil and Argentina. Eastern Atlantic: possibly England, also France and Bay of Biscay, to Gibraltar, Mediterranean Sea (including the Sea of Marmara, and excluding the Black Sea), Madeira, Canary Islands, Senegal, Gambia, Ghana, possibly Zaire, Angola, Namibia, South Africa (Northern and Western Cape Provinces); also Gough Island. Indo-West Pacific: South Africa (Eastern Cape and KwaZulu-Natal Provinces), Mozambique, Tanzania, Zanzibar, Kenya, Seychelles, Madagascar, Réunion, Mauritius, possibly Red Sea and Persian Gulf (Kuwait?), Sri Lanka, possibly Indonesia, Australia (Queensland, New South Wales, Victoria, Tasmania, South and Western Australia), New Zealand (including Norfolk, Stewart, and Chatham Islands), New Caledonia, Philippines (Mindanao, Palawan), China, Taiwan (Province of China), Japan, North Korea, South Korea, Russia (Siberia, possibly Sea of Okhotsk and Bering Sea), Bonin Islands (Tanna Island). Central Pacific: Marshall Islands, Hawaiian Islands, open ocean between Polynesia and South America. Eastern Pacific: Bering Sea and Gulf of Alaska to Gulf of California, including Canada (British Columbia) and the entire Pacific coast of the United States (Alaska, Washington, Oregon, California), and Mexico, also Panama, Ecuador, Peru, Chile, and Galapagos Islands (Compagno, 2001; Kabasakal, 2003; Zuffa *et al.*, 2002).

The great white shark is typically a rare animal, even though it has a very wide range. However, this predator is relatively abundant in the waters of South Africa, Central California, New England and South Australia (De Maddalena, 2007).

Habitat

The great white shark has one of the widest habitat ranges of any fish (Compagno, 2001). It prefers coastal and offshore waters of the continental and insular shelves in temperate seas, but regularly occurs in the offshore epipelagic zone (the illuminated surface zone, from the surface down to around 50 m, where there is enough light for photosynthesis and where living organisms concentrate). This shark is able to travel extensively, and can make transoceanic migrations and is known for tolerating temperature extremes, from the cold Bering Sea and sub–Antarctic islands to the warm inshore tropics (Compagno, 2001; Bonfil *et al.*, 2005). White sharks occasionally come close to shore, often in zones where the bottom drops very rapidly and prey are abundant (De Maddalena, 2009). This predator frequents the area close to banks, continental and oceanic islands, straits, and channels where it can easily find prey (Compagno, 2001; De Maddalena, 2002). In the tropics, it is occasionally sighted around coral reefs. The great white shark also penetrates shallow bays, estuaries and the intertidal zone in continental coastal waters, but is not known to occur in fresh water (Compagno, 2001). The great white shark can be found in depths ranging from the surface to at least 1280 m (Compagno, 1984). It is often found close to or at the surface in temperate waters, where it can be seen cruising slowly with its first dorsal fin and caudal fin upper lobe

The great white shark is often found close to or at the surface of temperate waters. Most of the time, they prefer a depth range of 13.9 m to 32.1 m (courtesy Joost Wenderich).

out of the water (Compagno, 1984). However, most of the time, it prefers a depth range of 13.9 m to 32.1 m (Goldman *et al.*, 1996).

Movements

Great white sharks are nomadic and may spend relatively short periods of time at a given site. These sharks may exhibit philopatry or site fidelity, which is the tendency to return to a specific location in order to breed or feed. The word philopatry is derived from the Greek, meaning "home-loving," although philopatry can manifest itself in several ways and can be applied to more than just the area where the animal was born. At least some individuals revisit these sites periodically, from a few days to over several years (Compagno, 2001).

Pronounced periodicity in white shark abundance may occur in some areas, apparently correlated with temperature and to some extent with life stage, or by movements of individuals or groups in response to prey concentrations. Also, shifts may occur in size and gender composition of white sharks off "white shark sites" such as fish banks or seal colonies where the sharks congregate (Compagno, 2001). In colder, higher latitudes at the periphery of its range in North America, the white shark moves into more northern areas when water masses warm up in the summertime (Compagno, 1984; Leim and Scott, 1966). In Monterey Bay, California, white sharks are present year round, but are slightly more common when water temperatures rise to 14 or 15°C than when it is below 11°C (Compagno, 2001).

Tagging and photo-identification of great white sharks have shown long-range movements. These movements include travel between northern and southern California (over 700 km), travel between Dyer Island and Mossel Bay, South Africa (about 300 km), from Dyer Island to KwaZulu-Natal, South Africa (over 1,100 km), from Mossel Bay to Delagoa Bay, Mozambique (about 1,500 km), and, most remarkably, from Dyer Island, South Africa, to the Exmouth Gulf of Western Australia (about 11,100 km) and back (Anderson and Goldman, 1996; Bonfil *et al.*, 2005; Compagno, 2001).

Like other sharks, great white sharks use the ambient electric and magnetic fields for orientation. In order to interpret the surface electric fields in terms of drift with respect to deeper water layers or to the ocean floor, the sharks should occasionally explore the depth profile of the electric field, as they may do during their deep dives (Carey and Scharold, 1990).

Ecological Role in Marine Communities

Great white sharks are apex predators, and we can imagine them at the top of an imaginary pyramid, called the pyramid of biomass. This pyramid represents the total

Great white sharks, as apex predators, are at the top of the pyramid of biomass (photograph by Callaghan Fritz-Cope, courtesy Pelagic Shark Research Foundation).

amount of energy and living mass in an ecosystem. The greatest amount of energy and biomass is present at the base of the pyramid with the producers (autotrophs), while the least amount of energy and biomass is found at the top of the pyramid with the highest level consumers, or apex predators. As the pyramid is ascended, energy decreases steeply, and so does the biomass that can be supported at each level. The smallest organisms are very numerous, and supply nourishment to the next largest organisms at the next level. As we move up this pyramid, organisms increase in size and decrease in quantity. Great white sharks play an important ecological role in marine communities. As predators, they are fundamental instruments of natural selection. As scavengers they process organic material so that it can then be used by autotrophs and heterotrophs at the bottom of the pyramid. Great white sharks influence the composition of marine ecosystems, contribute to ecosystem stability, and maintain biodiversity. These fish play a very significant role in the marine food web (De Maddalena, 2008).

Rate of Food Consumption

Great white sharks consume a relatively small amount of food. It has been estimated that a 4.6 m great white shark could survive 1.5 months between meals. It has also been estimated that about 30 kilograms of fatty tissue taken from a whale or a pinniped would provide enough energy to sustain a great white shark for this period (Carey *et al.*, 1982). Great white sharks feed intensively for a short time, and then feed very little for a longer period of time. The short period of feeding activity occurs when sharks are in a feeding mode, which is followed by a longer period of digestion when feeding is minimal.

Sharks expend considerable time and energy looking for prey and in hunting activity, so it is beneficial for a large shark to consume one large food item rather than to capture numerous small items. This notion is supported by the preference for large prey, such as tuna, other sharks, pinnipeds, dolphins, and marine turtles, which are dominant parts of the white shark diet (see page 52, "Diet"). A large cetacean carcass contains enough energy to sustain numerous great white sharks for a long period of time. The nutritional benefit a great white shark receives from a whole whale is enormous. Great white sharks seem to feed selectively on the energy-rich blubber layer of a cetacean carcass (Pratt *et al.*, 1982; Long and Jones, 1996).

Rate of consumption is estimated based primarily on the amount of food found in shark stomachs. Great white sharks that have empty stomachs have probably eaten at least 24 hours prior to capture. These sharks are able to evert their stomachs and expel the contents (see page 29, "Digestive System"), therefore, an empty stomach can't be considered proof of fasting. Food consumption needs to compensate for energy expended in growth and activities, including normal swimming, prey search, preda-

It has been estimated that a 4.6 m great white shark could survive 1.5 months between meals. It has also been estimated that about 30 kilograms of fatty tissue taken from a whale or a pinniped would provide enough energy to sustain a great white shark of similar size for this period (photograph courtesy Walter Heim).

tion and reproduction. Food consumption is related to the activity level of a particular species, which also relates to the rate at which food is digested.

The very active warm-blooded great white sharks have a higher metabolism rate than many other sharks that are cold-blooded animals. So, species phyletically far from the great white shark, such as the blue shark *Prionace glauca* (Linnaeus, 1758) and the sandbar shark *Carcharhinus plumbeus* (Nardo, 1827), are cold-blooded and eat 0.2 percent to 0.6 percent of their body weight per day. A species closer to the great white shark, such as the shortfin mako *Isurus oxyrinchus* Rafinesque, 1810, is warm-blooded and eats 3.0 percent of its body weight per day (Stillwell, 1991). A 1.5 m great white shark in captivity ate 1.6 percent of its body weight per day (Ezcurra *et al.*, 1996).

Diet

The great white shark is one of the more omnivorous sharks, and, in fact, they eat almost anything, though they are primarily carnivorous, or flesh-eating animals. It has been suggested that the great white shark favors an energy-rich fatty diet. In fact, these sharks show preference for northern elephant seals *Mirounga angustirostris* (Gill, 1866) and harbor seals *Phoca vitulina* (Linnaeus, 1758) over sea lions (family Otariidae) and sea otters *Enhydra lutris* (Linneaus, 1758), for pinniped pups over adults, and for baleen whale blubber over muscle (Pratt *et al.*, 1982; Long and Jones, 1996). Great white sharks are opportunistic feeders, meaning they are versatile and able to utilize diverse food sources depending on the availability of each food type. This species obtains most of its prey by killing it, but readily scavenges available carrion. Its diet in different areas may vary according to the availability and vulnerability of suitable prey (Compagno, 2001). Great white sharks usually prey on animals of smaller size. Live prey of great white sharks range in size from small schooling fishes to grey whale calves, *Eschrichtius robustus* (Lilljeborg, 1861). The diet of a great white shark is closely related to its size and age. As the size of the great white shark increases, the prey spectrum also increases as the larger shark can feed on larger prey. Larger white sharks (above 3 m in length) tend to prey more readily on marine mammals, and in general will feed on any large size prey. Sharks below 3 m in length feed more readily on small and medium-sized bony fishes and other sharks (Compagno, 2001).

Great white sharks feed on a wide variety of prey, including bony fishes, elasmobranchs, marine mammals, molluscs, crustaceans, sea turtles and birds. The bony fish and cartilaginous fish prey of the white shark includes a variety of sizes from small to large, and includes pelagic, demersal and benthic species. Great white sharks are known to congregate near concentrations of schooling bony fishes such as tuna, pilchards and bluefish (Compagno, 2001; De Maddalena, 2009). The great white shark's diet even includes a number of other shark species. Many sharks can eat conspecifics (De Maddalena, 2008). While large white sharks can attack hooked or injured conspecifics, and

The stomach of this 440 cm male great white shark caught on September 1, 1950, off Willapa Harbor, Washington, contained four salmon *Oncorhynchus* sp., vertebral columns of North Pacific hake *Merluccius productus*, rockfish *Sebastes* sp., the hides of two harbor seals *Phoca vitulina*, and 150 crabs including dungeness crabs *Cancer magister* and rock crabs *C. productus*. The set of jaws from this specimen is preserved at the University of Washington Fish Collection in Seattle, Washington, with cat. no. UW 044031 (photograph by Jeff Benca, courtesy University of Washington Fish Collection, Seattle).

can deliver severe bites to other healthy white sharks, no white shark has been found in another white shark's stomach. It has been suggested that white sharks have behavioral inhibitions on cannibalization under ordinary circumstances, as reflected from social interactions of white sharks within aggregations (Compagno, 2001) (see page 66, "Competition"). Marine mammals, both pinnipeds and cetaceans, are an important food source for white sharks (Compagno, 2001; De Maddalena, 2009). Pinnipeds tend to congregate in colonies where they are highly vulnerable (Compagno, 2001). Live small and medium-sized cetaceans, as well as dead large cetaceans, contribute a significant amount to the white shark diet (De Maddalena, 2009; Long and Jones, 1996). Sea turtles are occasionally eaten by the great white shark. Marine birds are commonly grabbed, killed and often eaten. Terrestrial mammalian carrion from slaughterhouses and other sources has also been found in the stomachs of white sharks. Invertebrate prey, including squids, abalone and other gastropods, bivalves, and crabs are also eaten (Compagno 2001). The occurrence of algae and seaweed in shark stomachs is usually associated with benthic animal remains. Sometimes the prey grasps the substratum when trying to escape, and the predator ingests both prey and algae (De Maddalena, 2008). In other circumstances, the prey may be hiding in the canopy of large algae, such as kelp *macrosystis* sp., and the algae is ingested with the hiding ani-

Live small and medium-sized cetaceans (shown here is a bottlenose dolphin *Tursiops truncatus*), as well as dead large cetaceans, contribute a significant amount to the white shark diet (photograph courtesy Walter Heim).

mal. Otters, sea lions and harbor seals are commonly found in the canopies of kelp forests.

The stomachs of some great white sharks contain inedible items and other oddities. These materials are rarely found in the stomachs of most shark species. However, the great white shark, tiger shark *Galeocerdo cuvier* (Peron & LeSueur, 1822) and the bull shark *Carcharhinus leucas* (Valenciennes, 1839), swallow inedible items more frequently than others. The list of oddities found in the stomachs of great white sharks from various locations (including the Mediterranean Sea, South Africa, Australia and other locations) includes stones of up to 7 kg, a 2.74 m long wire, a buoy chain, thirty-one hooks that were 15 cm long, pants, boots, shoes, baskets, a small board of cork, a raincoat, two or three coats, other clothes, a wig, a broom handle, an automobile license plate, a plastic bin, plastic bags, garbage, a sheet of cardboard, a duster, a ship scraper, a wicker-covered scent bottle, plastic bottles and two pumpkins (De Maddalena, 2008, 2009; Lineaweaver and Backus, 1969). Some researchers have hypothesized that stones, pebbles and other dense inedible items may be useful as ballast. Some of these materials are also thought to have been ingested as part of the stomach contents of the prey (De Maddalena, 2008).

A complete list of food items of the great white shark including both animals that are preyed upon live and those eaten when they are found dead, includes the following item (Black, 2008; Compagno, 1984, 2001; Cliff *et al.*, 1989; De Maddalena, 2009; De Maddalena and Révelart, 2008; Wilbur Irving Follett, unpublished data; Martin, 2003; Postel, 1958):

Porifera
    Sponges
Plants:
    Algae, eelgrass *Zoostera* sp.
Molluscs:
    Abalone *Haliotis* spp.
    Frog snail *Bursa californica* (Hinds, 1843)
    Snails (Class Gastropoda)
    Bivalves (Class Bivalvia)
    Squids (Order Teuthoidea)
    Cuttlefishes (family Sepiidae)
Crustaceans:
    Pacific rock crab *Cancer antennarius* (Stimpson, 1856)
    Dungeness crab *Cancer magister* (Dana, 1852)
    Red rock crab *Cancer productus* (Randall, 1839)
Echinoderms:
    Sea stars *Asteris* sp.
Bony fish:
    Green sturgeon *Acipenser medirostris* Ayres, 1854

Atlantic menhaden *Brevoortia tyrannus* (Latrobe, 1802)
South American pilchard *Sardinops sagax* (Jenyns, 1842)
European pilchard *Sardina pilchardus* (Walbaum, 1792)
Anchovies (family Engraulidae)
Chinook salmon *Oncorhynchus tshawytscha* (Walbaum, 1792)
Pacific salmon *Oncorhynchus* spp.
Sea catfishes (family Ariidae)
Searobins *Prionotus* spp.
North Pacific hake *Merluccius productus* (Ayres, 1855)
Silver hake *Merluccius bilinearis* (Mitchill, 1814)
Hakes *Urophycis* spp.
Lings (family Gadidae)
Flounders and halibuts (family Paralichthyidae and family Pleuronectidae)
Barracudas (family Sphyraenidae)
Santer seabream *Cheimerus nufar* (Valenciennes, 1830)
English seabream *Chrysoblephus anglicus* (Gilchrist & Thompson, 1908)
Slinger seabream *Chrysoblephus puniceus* (Gilchrist & Thompson, 1908)
Smallspotted grunter *Pomadasys commersonni* (Lacepède, 1801)
Natal pandora *Pargellus natalensis* Steindachner, 1903
Hapuka *Polyprion oxygeneios* (Schneider & Forster, 1801)
Bluefish *Pomatomus saltatrix* (Linnaeus, 1766)
Butterfishes (family Stromateidae)
Striped bass *Morone saxatilis* (Walbaum, 1792)
White weakfish *Atractoscion nobilis* (Ayres, 1860)
Jacks (family Carangidae)
Chub mackerel *Scomber japonicus* (Houttuyn, 1782)
Northern bluefin tuna *Thunnus thynnus* (Linnaeus, 1758)
Albacore *Thunnus alalunga* (Bonnaterre, 1788)
Yellowfin tuna *Thunnus albacares* (Bonnaterre, 1788)
Atlantic bonito *Sarda sarda* (Bloch, 1793)
Bullet tuna *Auxis rochei* (Risso, 1810)
Frigate tuna *Auxis thazard* (Lacepède, 1800)
Swordfish *Xiphias gladius* (Linnaeus, 1758)
Western Australia salmon *Arripis truttacea* (Cuvier, 1829)
Squirefish *Pagrus auratus* (Forster, 1801)
Madagascar meagre *Argyrosomus hololepidotus* (Lacepède, 1801)
Scorpionfish *Scorpaena* sp.
Black rockfish *Sebastes melanops* (Girard, 1856)
Rockfishes *Sebastes* spp.
Cabezon *Scorpaenichthys marmoratus* (Ayres, 1854)
Lingcod *Ophiodon elongatus* (Girard, 1854)

1 — Biology, Ethology and Ecology of the Great White Shark

    Mullet *Mugil* sp.
    Common dentex *Dentex dentex* (Linnaeus, 1758)
Cartilaginous fish:
    Broadnose sevengill shark *Notorynchus cepedianus* (Péron, 1807)
    Spiny dogfish *Squalus acanthias* (Linnaeus, 1758)
    Whale shark *Rhincodon typus* (Smith, 1828)
    Basking shark *Cetorhinus maximus* (Gunnerus, 1765)
    Thresher shark *Alopias* sp.
    Sandtiger shark *Carcharias taurus* (Rafinesque, 1810)
    Shortfin mako *Isurus oxyrinchus* (Rafinesque, 1810)
    Grey smooth-hound shark *Mustelus californicus* (Gill, 1864)
    Dusky smooth-hound *Mustelus canis* (Mitchill, 1815)
    Brown smooth-hound *Mustelus henlei* (Gill, 1863)
    Tope shark *Galeorhinus galeus* (Linnaeus, 1758)
    Milk shark *Rhizoprionodon acutus* (Rüppell, 1837)
    Bronze whaler *Carcharhinus brachyurus* (Günther, 1870)
    Dusky shark *Carcharhinus obscurus* (Lesueur, 1818)
    Sandbar shark *Carcharhinus plumbeus* (Nardo, 1827)
    Blue shark *Prionace glauca* (Linnaeus, 1758)
    Scalloped hammerhead *Sphyrna lewini* (Griffith & Smith, 1834)
    Lesser guitarfish *Rhinobatus annulatus* (Müller & Henle, 1841)
    Giant guitarfish *Rhynchobatus djiddensis* (Forsskål, 1775)
    Stingrays *Dasyatis* spp.
    Bat eagle ray *Myliobatis californica* Gill, 1865
    Bull ray *Pteromylaeus vinus* (Geoffroy Saint-Hilaire, 1817)
    Spotted ratfish *Hydrolagus colliei* (Lay & Bennett, 1839)
    Plownose chimaeras (family Callorhynchidae)
Reptiles:
    Leatherback sea turtle *Dermochelys coriacea* (Vandelli, 1761)
    Green sea turtle *Chelonia mydas* (Linnaeus, 1758)
    Loggerhead sea turtle *Caretta caretta* (Linnaeus, 1758)
Birds:
    Jackass penguin *Spheniscus demersus* (Linnaeus, 1758)
    Cape cormorant *Phalacrocorax capensis* (Sparrman, 1789)
    Gannets *Sula* spp.
    Grey-headed gull *Larus cirrocephalus* Vieillot, 1818
    Gulls (family Laridae)
    Giant petrels *Macronectes* sp.
    Pelicans (family Pelicanidae)
Pinnipeds:
    Harbor seal *Phoca vitulina* (Linnaeus, 1758)

Grey seal *Halichoerus grypus* (Fabricius, 1791)
Northern elephant seal *Mirounga angustirostris* (Gill, 1866)
Mediterranean monk seal *Monachus monachus* (Hermann, 1779)
Australian sea lion *Neophoca cinerea* (Peron, 1816)
New Zealand sea lion *Phocarctos hookeri* (Gray, 1844)
California sea lion *Zalophus californianus* (Lesson, 1828)
Steller sea lion *Eumetopias jubatus* (Schreber, 1776)
South American fur seal *Arctocephalus australis* (Zimmerman, 1783)
Guadalupe fur seal *Arctocephalus townsendi* (Merriam, 1897)
Australian fur seal *Arctocephalus pusillus doriferus* (Wood Jones, 1925)
Cape fur seal *Arctocephalus pusillus pusillus* (Schreber, 1775)
New Zealand fur seal *Arctocephalus forsteri* (Lesson, 1828)

Cetaceans:
Harbor porpoise *Phocoena phocoena* (Linnaeus, 1758)
Dall's porpoise *Phocoenoides dalli* (True, 1885)
Pacific white-sided dolphin *Lagenorhynchus obliquidens* (Gill, 1865)
Dusky dolphin *Lagenorhynchus obscurus* (Gray, 1828)
Common dolphin *Delphinus delphis* (Linnaeus, 1758)
Bottlenose dolphin *Tursiops truncatus* (Montagu, 1821)
Striped dolphin *Stenella coeruleoalba* (Meyen, 1833)
Risso's dolphin *Grampus griseus* (Cuvier, 1812)
Dwarf sperm whale *Kogia simus* (Owen, 1866)
Pygmy sperm whale *Kogia breviceps* (de Blainville, 1838)
Sperm whale *Physeter macrocephalus* (Linnaeus, 1758)
Indo-Pacific humpbacked dolphin *Sousa plumbea* (G. Cuvier, 1829)
Stejneger's beaked whale *Mesoplodon stejnegeri* (True, 1885)
Cuvier's beaked whale *Ziphius cavirostris* (Cuvier, 1823)
Grey whale *Eschrichtius robustus* (Lilljeborg, 1861)
Fin whale *Balaenoptera physalus* (Linnaeus, 1758)
Blue whale *Balaenoptera musculus* (Linnaeus, 1758)
Humpback whale *Megaptera novaeangliae* (Borowski, 1781)

Terrestrial mammals:
Human *Homo sapiens sapiens* (Linnaeus, 1758)
Dog *Canis lupus familiaris* (Linnaeus, 1758)
Cat *Felis catus* (Linnaeus, 1758)
Domestic sheep *Ovis aries* (Linnaeus, 1758)
Domestic goat *Capra aegagrus hircus* (Linnaeus, 1758)
Horse *Equus caballus* (Linnaeus, 1758)
Domestic pig *Sus scrofa scrofa* (Linnaeus, 1758)
Cattle *Bos* spp.

Predatory Tactics

Great white sharks need to be able to capture prey as efficiently as possible, and behavioral adaptations allow these predators to catch and feed efficiently. Since evolution depends upon survival of the fittest, predatory success plays a very important role in great white shark survival. These formidable predators have a wide variety of predatory strategies. Great white sharks are versatile predators, and do not rely exclusively on a single tactic to capture prey (De Maddalena, 2008).

It is thought that the great white shark would have difficulty capturing a healthy, fast-swimming prey item if the animal were aware of the predator's presence. The great white shark is a stalking hunter, thus the success of this predator depends on both speed and the element of surprise (Strong, 1996b). Great white sharks attack fast-swimming animals on which they feed suddenly and violently. The victim typically never sees the shark until it is too late and is overwhelmed by the unexpected assault and the violent force with which it is executed. These sharks have been reported to attack seals, sea lions, sea otters, dolphin, tuna, and humans using this method (Miller and Collier, 1980; Tricas and McCosker, 1984; Long and Jones, 1996; Long *et al.*, 1996; Ames *et al.*, 1996; Burgess and Callahan, 1996; Levine, 1996; West, 1996).

In great white sharks, the tips of the underside of the pectoral fins are black (see page 23, "Color"). It has been suggested that this coloration is a fine tuning of the camouflage pattern. The black tips compensate for the flash of white that might otherwise alert potential prey when the fins flex as the predator turns (Ellis and McCosker, 1991).

Gabriotti and De Maddalena (2004) described a particular approach on possible prey performed by great white sharks at the Neptune Islands, in the mouth of Spencer Gulf, South Australia. The large quantities of tuna oil, blood and macerated tuna used to attract the sharks to the vessel for shark cage diving also attracted schools of jack mackerel *Trachurus declivis* (Jenyns, 1841). Some great white sharks, of estimated total lengths of 3.5–4 m, performed the following tactic when approaching the observer. On the first pass the shark swam towards the cage, approaching from the side of the cage that was covered by numerous jack mackerels. Then the shark observed the divers, remaining at the visibility limit behind the school of fish. Gabriotti and De Maddalena (2004) have suggested that great white sharks may use this tactic to increase the element of surprise when attacking prey, and to estimate the size of the prey, and its strength, before executing an attack. As the white shark moves toward a prey item, it uses a tightly packed school of fish as camouflage to obscure its approach. The shark can continue to evaluate the parameters of the predatory situation and remain undetected at the visibility limit behind the school of fish, swimming on an axis aligning the shark, school of fish, and prey. Gabriotti and De Maddalena (2004) named this behavior "hidden approach." The observation that larger white sharks did not show this kind of approach behavior could be related to their greater size and strength. It

It has been suggested that in the great white shark, the black tips of the underside of the pectoral fins compensate for the flash of white that might otherwise alert a potential prey when the fins flex as the predator turns. This replica cast is made from a 484 cm TL male great white shark caught east of Block Island, Connecticut, on August 5, 1983. It is on exhibit to the public at the Department of Marine Sciences of the University of Connecticut in Groton, Connecticut, without cat. no. (photograph courtesy J. Evan Ward, Department of Marine Sciences of the University of Connecticut, Groton).

could also be further indication of a lack of necessity for such a cautious behavior beyond a certain point in the growth of the shark.

Pinnipeds are a major food source for great white sharks (see page 52, "Diet"), so the hunting strategy of this predator is adapted to the life history of pinnipeds, which form colonies on islands and coastlines. These sharks carry out surprise attacks on sea lions and seals swimming at the sea surface.

Great white sharks approaching prey can be oriented horizontally or vertically. The predator uses its heavy mass and speed to violently ram and stun the prey. During vertical approaches, the great white shark attacks its prey from below, by swimming from depths as deep as 17 m and moving on a line that is 45°–90° oriented from the prey (Strong, 1996b). The shark swims with such extreme velocity that prey have been observed being propelled out of the water by the force of the shark's impact. Often the prey is disoriented, and therefore incapable of resistance. The majority of approaches observed during a study of white sharks in Spencer Gulf, Australia, were horizontal,

1 — Biology, Ethology and Ecology of the Great White Shark

but a smaller number of vertical approaches were also observed (Strong, 1996b). The predator is less visible when coming directly from below, and it has the best view of the prey silhouetted against the surface light. The prey has fewer escape paths as the prey is pinned against the surface (escape in the same direction as that of the predator is impossible) (Strong, 1996b). Great white shark attacks on humans can also be oriented horizontally or vertically. Usually the shark approaches from below, from the side or from behind its victim. Frontal attacks and attacks from above are rare (Burgess and Callahan, 1996; Levine, 1996; Miller and Collier, 1980). Strong (1996b) has also observed 2.2 m young white sharks vertical swimming, indicating that this behavior precedes graduation to larger prey. As stated previously, young white sharks measuring less than 3 m in length feed primarily on fish, while larger individuals are important predators and scavengers of marine mammals (see page 52, "Diet").

Several studies on the hunting strategy of great white sharks have been conducted at Southeast Farallon Islands off San Francisco, California, Año Nuevo Island off Santa Cruz, California, and Seal Island, South Africa (Ainley *et al.*, 1981, 1985; Klimley *et al.*, 1996, 2001; Martin *et al.*, 2005). These islands and the adjacent coastline are home to

A breach, where the great white shark leaps partially or completely out of the water, is a commonly employed predatory tactic (photograph courtesy Gaspare Schillaci).

colonies of phocids and otariids, including California sea lions *Zalophus californianus*, northern elephant seals *Mirounga angustirostris*, Steller sea lions *Eumetopias jubatus*, harbor seals *Phoca vitulina*, and Cape fur seals *Arctocephalus pusillus pusillus*.

The prey killed by a shark is not always identifiable owing to the fact that dead sea lions sink, while dead seals float. Nevertheless, seals are eaten more often than sea lions by great white sharks. This preference may reflect the fact that sea lions, with their speed and maneuverability, are more difficult to catch, but there are other differences that may influence differential pinniped predation. Sea lions often occur in groups, while elephant seals are usually solitary when away from shore. The solitary behavior of elephant seals makes them more vulnerable to predation. When sea lions swim, they use their front flippers to push themselves through the water, while their rear flippers are used to help steer. Seals swim by flapping their rear flippers, while their front flippers help them steer (Ainley *et al.*, 1985; Ellis & McCosker, 1991; Long *et al.*, 1996).

Great white shark attacks on elephant seals have been repeatedly observed, and the attacks follow a general pattern (Long *et al.*, 1996). The great white shark executes an initial attack on the near-surface pinniped from below or behind, usually inflicting a deep bite on the rear part of the body. Biting the rear part of the elephant seal disables the propulsive mechanism of the pinniped. The initial attack is often followed by a waiting period in which the elephant seal does not usually attempt to flee. Reasons for the lack of evasion include the disabling nature of the initial wound, and the onset of shock due to extreme blood loss. The elephant seal expires due to the loss of blood, and the white shark returns within 1 to 5 minutes to begin consuming the dead animal. This predatory tactic enables the shark to obtain its meal with minimal risk of injury as well as minimal energy expenditure. This behavior has been termed "bite-and-spit" (Tricas and McCosker, 1984). In some cases, however, the prey may escape, and survive or die far from the attack location if it is strong enough to swim. Sometimes pinnipeds drag themselves out of the water and die on the beach as a result of shark bites. "Bite and spit" behavior may allow many human victims the possibility to escape or be rescued. However, West (1996) found that a high percentage of great white shark attacks on humans do not conform to this behavior; consequently "bite-and-spit" is not the rule. Even biting the hindquarters is not a rule since Long *et al.* (1996) observed wounded and dead elephant seals with shark bite scars distributed over the entire body, including the head and neck. In fact, sometimes great white sharks attempt to decapitate their prey. Sea lions that survive a great white shark attack have bite scars distributed predominantly on the posterior part of the body. Sea lions are often able to escape from a great white shark attack because a bite to their rear flippers does not incapacitate them. The great white shark executes a successful attack on a sea lion usually by inflicting a bite on the mid part of the body (Ainley *et al.*, 1981, 1985; Long *et al.*, 1996).

At Seal Island, South Africa, most attacks lasted less than one minute and consisted of a single breach. A polaris breach, where the shark leaps partially or completely

1 — Biology, Ethology and Ecology of the Great White Shark

out of the water in a vertical or nearly vertical head-up orientation, was the most commonly employed initial strike. A surface lunge where the shark, oriented dorsum up and with its back partially out of the water, accelerates quickly with its jaws held open, exposing the upper teeth, was the most frequent second event. This was followed by a lateral snap, which is protruding the upper jaw and grasping the prey with the anterolateral teeth, with the head flexed sideways toward the prey. Great white sharks at Seal Island bite prey obliquely using their anterolateral teeth via a sudden lateral snap of the jaws and not perpendicularly with their anterior teeth. Analysis of white shark upper tooth morphology and spacing suggest the reversed intermediate teeth of white sharks occur at the strongest part of the jaw and produce the largest wound (Martin *et al.*, 2005).

Young pinnipeds are attacked more frequently than adults. White sharks appear to prefer young elephant seals, in the range of 1 to 2 years of age, and Cape fur seals up to a year old (Martin *et al.*, 2005; Michael, 1993; Long *et al.*, 1996). At the Farallon Islands, great white sharks kill more seals in the fall. Similarly, at Año Nuevo Island, shark-bitten seals are more common during the fall and winter. At this time, young elephant seals are more abundant at these sites (Long *et al.*, 2006). Young pinnipeds

As in the case of this California sea lion *Zalophus californianus*, sea lions that survive a great white shark attack have bite scars distributed predominantly on the posterior part of the body (photograph courtesy Walter Heim).

are the target of attacks more frequently because they have a higher fat content, are less vigilant, have less experience with sharks than do adults, and may behave in a manner that makes them more vulnerable to predation (Michael, 1993). In fact, immature pinnipeds are observed with great white shark bite scars more often than adults.

The great white shark prefers coastal waters with a median depth of 20 meters because the density of juvenile seals is highest close to shore. The area patrolled by the great white sharks at Southeast Farallon Islands and Año Nuevo Island extends from a few meters to 1.3 km offshore and surrounds the pinniped colonies, where they would have the greatest chance of capturing prey. Within this area, predation is most frequent adjacent to rookeries and beaches where pinniped colonies concentrate, and particularly near entry and departure points of pinnipeds. Great white sharks tagged and monitored at Año Nuevo Island rarely ventured far from shore, but stayed very close and at times approached to within two meters of the shore. At Año Nuevo Island, the sharks visited the seal colony area every day and patrolled this zone both during the daytime and at night. Some of the sharks moved back and forth parallel to the shoreline, 200–300 meters from shore, where they were ideally positioned to intercept and stalk seals and sea lions departing from and returning to the rookeries (Klimley *et al.*, 2001). Predation by the great white shark on pinnipeds at South Farallon Islands is more frequent during higher tides than during lower tides, because northern elephant seals are more numerous in the water during higher tides. The South Farallon Islands pinniped population has reached carrying capacity, and space at high tide is scarce. Consequently, during higher tides, juvenile pinnipeds are forced to enter the water owing to lack of space (Anderson *et al.*, 1996). Other environmental factors, including water clarity, weather, lunar illumination and sea temperature, may affect frequency of predatory activities (Pyle *et al.*, 1996). Most sharks have crepuscular (the low-light transition times between night and day), or nocturnal feeding habits, feeding primarily at twilight or in darkness (De Maddalena, 2008). Under the cover of complete darkness, prey have more difficulty detecting predator approach. The great white shark is equally active during the day and night (Klimley *et al.*, 2001). However, success of great white shark predation on Cape fur seals *Arctocephalus pusillus pusillus* at Seal Island, South Africa, is greatest within one hour of sunrise and decreases rapidly with increasing ambient light. Active predation on seals ceases when the success rate drops to ±40 percent (Martin *et al.*, 2005).

Some great white sharks have been reported to incapacitate prey by biting off their tails. A vertical approach may be commonly used by these predators to obtain this result when pursuing fast-swimming animals, such as tuna and dolphin. A single bite on the caudal peduncle of these large prey can sever swimming muscles, the spinal column and blood vessels, thereby immobilizing the prey. In these cases, the great white shark swims at a greater depth than its prey, giving them a view of what is above and a higher probability of seeing and attacking the prey, via a rapid vertical approach, before being seen themselves. This predatory tactic explains the presence of wounds over the posterior part of the body and caudal peduncle inflicted on some bluefin tuna and dolphins by great white sharks (Arnold, 1972; De Maddalena, 2008; Ellis and McCosker, 1991).

In order to avoid detection by dolphins and porpoises, great white sharks approach these mammals from below, above or behind, because odontocetes have an anteriorly directed sonar and a lateral visual field. Long and Jones (1996) examined numerous small cetaceans with wounds on the caudal peduncle, urogenital region, abdominal area, and dorsum, while wounds to the head and flanks were less common. Urogenital and abdominal regions are vulnerable areas. Shark bite scars are more often observed on the dorsum of live cetaceans since this body region is less vulnerable, providing higher probabilities of surviving the attack (Celona *et al.*, 2006; Long and Jones, 1996).

In order to feed, the great white shark often lifts part of its body out of the water, commonly exposing the entire head and pectoral fins. These predators show a particular behavior not found in other sharks, called spy-hopping, in which they raise the head out of the water to investigate and locate potential prey. For example, this behavior allows the predator to locate pinnipeds resting on the rocks (Compagno, 1984; Martin, 2003).

Great white sharks often bite animals that are not consumed, on many occasions killing them in the process. This happens with humans as well as with sea otters *Enhydra lutris*, northern fur seals *Callorhinus ursinus* (Linnaeus, 1758) and African penguins *Spheniscus demersus*. Attacks on these animals by great white sharks have resulted in severe wounds and death for a substantial number of individuals. For example, Ames *et al.* (1996) have shown that sea otter mortality caused by great white sharks reaches 20 percent per year in the Año Nuevo Island area (Ames *et al.*, 1996). Collier *et al.* (1996) demonstrated that great white sharks also attack inanimate objects of a variety of shapes, sizes and colors, none resembling the shape, size or color of a pinniped. Consequently, they have suggested that these predators are in such cases determining the suitability of the prey as food. Martin (2003) and Collier (2003) suggested that many great white shark attacks may be a form of play or a hunting practice. Great white sharks learn from past experience and are able to refine their predatory skills, improving their hunting abilities (Martin, 2003).

Not only is the great white shark a predator of live animals, but it is also a scavenger (see page 52, "Diet"). An important distinction must be made between preying on a live animal and eating a dead animal. Scavengers feed on dead organisms. Large marine animal carcasses are known to attract great white sharks. Feeding on dead animals is very advantageous because it often provides enough energy to sustain a shark for long periods, with minimal energy expenditure (De Maddalena, 2008). Great white sharks are solitary hunters but may gather in relatively high numbers of up to ten individuals or more, at a large cetacean carcass (Compagno, 2001; De Maddalena, 2008).

Competition

Competition occurs when two or more sharks simultaneously attempt to feed on the same prey. Researchers who have studied shark behavior have observed both interspecific (between different species) and intraspecific (between the same species) competition. Access to food is often established through agonistic behavior, in which individuals of some species are very aggressive towards others (De Maddalena, 2008). Evidence indicates the existence of combat among great white sharks. Wounds that are the result of intraspecific aggression have been observed on the bodies of great white sharks, in both males and females, and in both mature and immature individuals. These wounds are often more severe than "love bites." In rare cases, agonistic behavior takes the form of real fighting. In fact, most scars resulting from intraspecific combat do not last more than a few years. These agonistic behaviors have a communication function, as the predator attempts to communicate with individuals of the same and other species using particular signals before attacking. These behaviors may function as a warning from a shark, telling other sharks that it is ready to attack and able to inflict severe injury. Agonistic behavior can prevent direct fights, forcing the other individual to flee without resorting to fighting (De Maddalena, 2008).

Most sharks are not highly social animals, but often form temporary social structures (De Maddalena, 2008). Great white sharks are sometimes found in small groups, but not tightly organized schools. Social hierarchies between great white sharks and other great white sharks, and between great white sharks and sharks of other species have been reported near a food source. Hierarchies serve as an anti-predatory tactic on the part of the subordinate shark (De Maddalena, 2008).

An order of dominance has been shown to exist based on species, size and gender. Great white sharks dominate blue sharks *Prionace glauca* when both species are feeding. According to Long and Jones (1996), blue sharks do not scavenge on a whale carcass when white sharks are feeding.

Social hierarchies among great white sharks have been reported, and are based on size. White sharks must possess a keen awareness of their own size. Competition is evident between great white sharks attracted to a food source. Smaller white sharks usually move away from larger members of their own species. In order to compare their relative size or to intimidate a conspecific, pairs of great white sharks have been observed swimming parallel to each other, until one of the two surrenders to the dominant individual and accelerates away. In other cases, a pair of great white sharks swim on a collision course and the subordinate individual gives way (Martin, 2003).

When a number of great white sharks are present around a carcass, only one or two individuals feed at a time. They seem to take turns feeding, and each individual feeds for about the same amount of time. In some cases, when two sharks feed simultaneously, they stay on opposite sides of the carcass. This behavior was described from observations of great white sharks feeding on carcasses of large cetaceans including a

1 — Biology, Ethology and Ecology of the Great White Shark

blue whale *Balaenoptera musculus* (Linnaeus, 1758), a fin whale *Balaenoptera physalus* (Linnaeus, 1758), and a humpback whale *Megaptera novaeangliae* (Borowski, 1781) (Long and Jones, 1996; Curtis et al., 2000). Many of the great white sharks seen near a whale carcass had lacerations that were probably inflicted by members of their own species while competing for the carcass.

Competition is even more evident among great white sharks attracted to a smaller animal carcass. In this situation sharks often contest ownership of the food item. When a great white shark kills prey such as a pinniped, other great white sharks are attracted to the site of the kill to feed. In these instances, competitors have to be dissuaded by prompt aggressive reaction. The first shark responds to these uninvited guests with threat displays indicative of competition for the prey. A "tail slap" involves a pair of sharks that lift their caudal fins and splash water at each other. An individual is permitted to eat the prey only if the vigor and frequency of its tail slap is greater than that of its competitor. A "breach" is when a shark leaps out of the water, two-thirds of the body emerging at an angle of 30°–60° to the sea surface, and is a less common behavior that may be a higher intensity display (Kilmley *et al.*, 1996). The great white shark can also perform other threat displays, such as gaping its lower jaw slightly, protruding the upper jaw (Martin, 2003; Strong, 1996a), or swimming with the back arched

Great white sharks have been observed to dominate blue sharks *Prionace glauca* when both species are feeding. Blue sharks do not scavenge on a whale carcass when white sharks are feeding (photograph courtesy Walter Heim).

and both pectoral fins lowered (Compagno, 2001). Another interesting great white shark behavior called "repetitive aerial gaping" has been described by Strong (1996a). When a great white shark is prevented from reaching a bait when the wrangler pulls it away, the predator tries to seize it by holding its head out of the water and rolling onto its side. It then opens and closes its mouth slowly, displaying partial gapes while swimming slowly at the surface. The repetitive aerial gaping is not oriented toward the food or other objects. Frustration can give rise to aggression, and under similar conditions great white sharks have been observed to bite conspecifics. Strong (1996a) has hypothesized that repetitive aerial gaping may be a manifestation of frustration, and may function to reduce intraspecific aggression, redirecting frustration and avoiding attacks on other great white sharks.

In general, it seems great white sharks do not possess any territory, and they do coexist with other members of their species without much conflict. They often show philopatry, or a special preference for an area where they stay or to which they return periodically (see page 49, "Movements"). In a study on great white sharks conducted in the waters surrounding pinniped colonies at Año Nuevo Island, California, researchers observed an almost complete absence of territoriality. Although each of the great white sharks spent more time in a slightly different location than the other individuals, and some sharks patrolled certain areas preferentially, all the individuals frequently moved over the same areas. There was no evidence that each individual defended an area as territory (Klimley *et al.*, 2001).

Predators and Parasites

Great white sharks are at the top of the pyramid of biomass (see page 49, "Ecological Role in Marine Communities"), and have few enemies. Only a few creatures prey on great white sharks. They include humans, and great white sharks themselves, but probably only under particular circumstances such as when another white shark is hooked or injured (see page 52, "Diet"). Killer whales *Orcinus orca* (Linnaeus, 1758) probably prey on great white sharks, but only in very rare cases with only a single case of predation documented (Pyle *et al.*, 1999).

Nevertheless, great white sharks are hosts to numerous parasites, such as copepods, trematods, and cestodes. These parasitic organisms live in or on the great white sharks in order to obtain sustenance from them. Multiple infections of great white sharks by parasites are common. Parasites can cause serious lesions, sometimes resulting in severe diseases (De Maddalena, 2008). Parasites of the great white shark include the copepods *Anthosoma crassum* (Abildgaard, 1794), *Echthrogaleus coleoptratus* (Guerin-Meneville, 1837), *E. denticulatus* (Smith, 1874), *Dinemoura producta* (Müller, 1785), *D. latifolia* (Steenstrup and Lutken, 1861), *Nesippus orientalis* (Heller, 1868), *Pandarus bicolor* (Leach, 1816), *P. smithii* (Rathbun, 1886), *P. satyrus* (Dana, 1852), *Achtheinus*

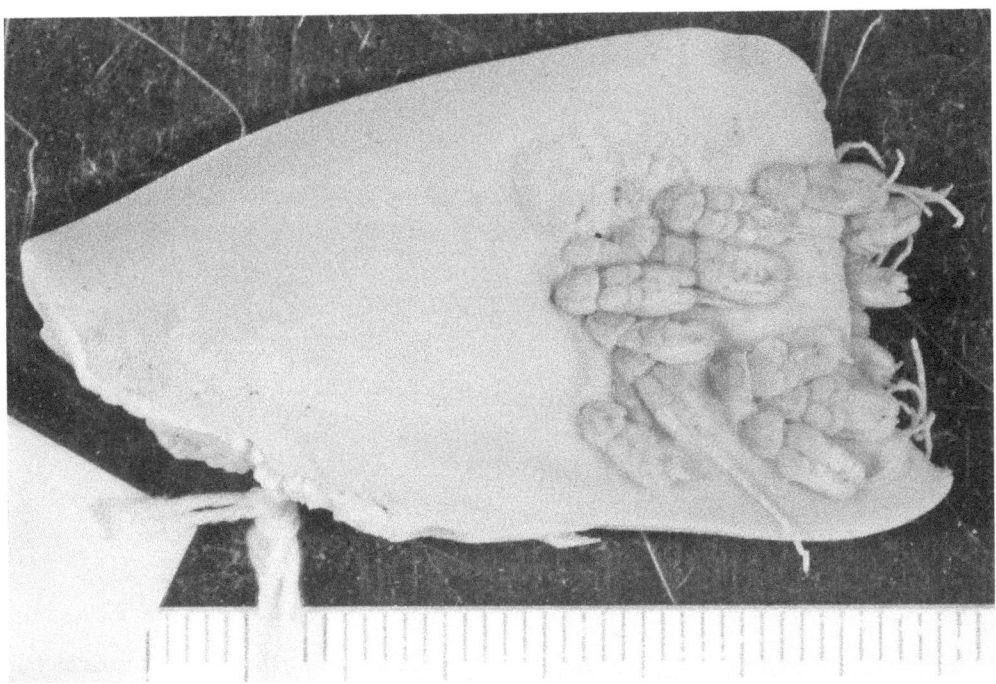

Pelvic fin with parasitic copepods from a male great white shark that was captured on November 15, 1945, off La Jolla, just north of Scripps Pier, California, U.S.A., preserved at the Museum of Comparative Zoology of Harvard University in Cambridge, Massachusetts, with cat. no. MCZ 36470 (photograph courtesy Andrew Williston, Museum of Comparative Zoology, Harvard University).

oblongus (Wilson, 1908), and *Nemesis lamna* (Risso, 1826), the cestodes *Tetrarhynchus megacephalus* (Rudolphi, 1819), *Dinobothrium septaria* (Van Beneden, 1889), *Clistobothrium carcharodoni* (Dailey and Vogelbein, 1990), and *Phyllobothrium loliginis* (Leidy, 1887), and the trematod *Distomum continuum* (Benz *et al.*, 2003; Brian, 1906; Dailey and Vogelbein, 1990; Hansson, 1998; Hewitt, 1967, 1979; Sumner *et al.*, 1913; Tortonese, 1956). Benz *et al.* (2003) proposed that the species-rich infections of some great white sharks may be the result of the wide geographical range of individual sharks through waters inhabited by other elasmobranchs.

Mutualism

Like all large sharks, great white sharks are sometimes accompanied by remoras and pilot fish. The remoras, including the common remora *Remora remora* (Linnaeus, 1758), live sharksucker *Echeneis naucrates* (Linnaeus, 1758) and white suckerfish *Remorina albescens* (Temminck and Schlegel, 1850), are bony fish of the family Echeneidae. Remoras have a dorsal suction disk formed from their modified dorsal fin, which they

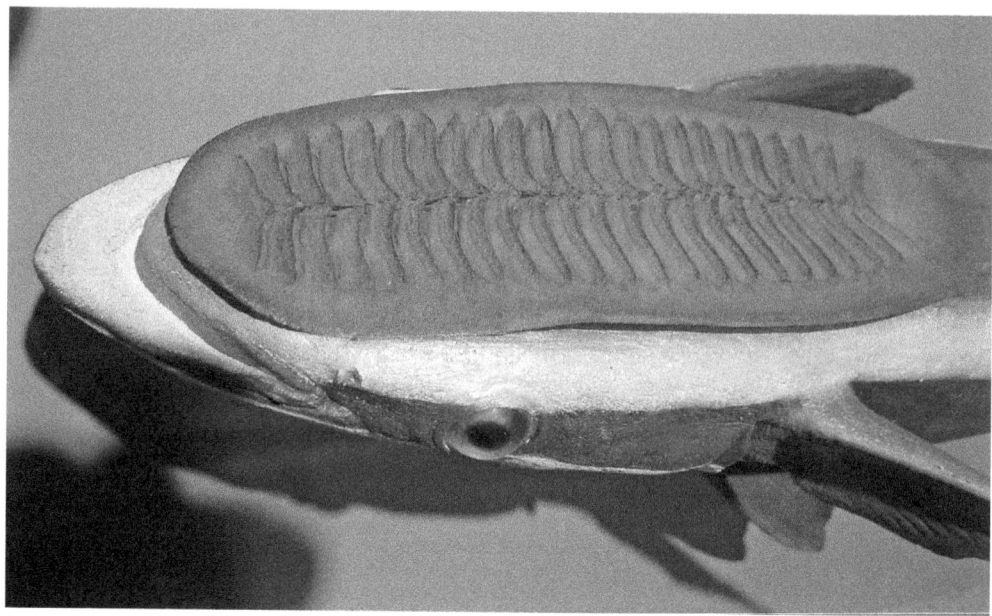

Like all large sharks, great white sharks are sometimes accompanied by remoras. In the photo, the dorsal suction disk of a live sharksucker *Echeneis naucrates* (photograph by Alessandro De Maddalena, courtesy Aquazoo-Löbbecke Museum, Düsseldorf, Germany).

use to attach themselves to sharks, mantas, marine turtles and other large creatures. They use this suction disk only when the large animal changes direction or slows down. The pilot fish *Naucrates ductor* (Linnaeus, 1758) is a bony fish of the family Carangidae that is often associated with cartilaginous fish, bony fish and marine turtles.

The relationships between pilot fish and great white sharks, and between remoras and great white sharks, are cases of mutualism, because both organisms benefit from each of these relationships. Pilot fish and remoras benefit from the relationship with a great white shark by eating the shark's scraps of food or excrement and parasites, as well as by riding the shark's bow wave. Consequently, great white sharks benefit from the relationship with remoras and pilot fish by being cleaned of parasites. Observations seem to indicate that pilot fish are much more common companions for great white sharks than the remoras, which are observed more sporadically on the body or in the proximity of these large predators (De Maddalena, 2009).

Attacks on Humans

People are occasionally killed or injured by sharks. Only three species are involved in most incidents, including the great white shark, tiger shark *Galeocerdo cuvier* and bull shark *Carcharhinus leucas*. The great white shark is responsible for the highest number of attacks (De Maddalena, 2008). Compared to the total number of deaths from any other form of water-related activity, deaths caused by great white sharks are very low. The fearsome reputation of these animals is exaggerated. Despite the large number of people who frequent beaches and seas, and the fact that great white sharks are present in almost all waters, great white shark attacks on man are extremely rare.

The Global Shark Attack File (GSAF) gathers data on shark attacks worldwide. It is maintained by the Shark Research Institute (SRI) based in Princeton, New Jersey. The first author of this book collaborates with the GSAF as a regional investigator for the Mediterranean Sea. The GSAF quantitative data is available to the medical profession, the scientific community, the media and the general public to provide accurate and current data on the subject of shark/human interactions. In addition, the file gathers medical data in a format that can be utilized by physicians and surgeons who are called upon to treat victims of shark attacks (De Maddalena *et al.*, 2007).

The actual number of great white shark attacks around the world is unknown. The incidents recorded represent only a portion of the actual total. In fact, in many areas of the world, shark attacks go unrecorded. In many Third World countries, shark attacks are rarely reported. In some countries, reports of attacks are intentionally suppressed in order to avoid extreme reactions from the media and damage to the tourism industry (De Maddalena, 2008). Moreover, only a portion of the attacks can be attributed to a particular species, because post-attack identification of the offending shark is often difficult. In most cases, the definitive identification of the species responsible for an attack can only be made if tooth fragments are found in the victim's wounds, or if the specimen is photographed or captured, or if a witness is a marine life expert (De Maddalena, 2008).

However, the reported data is confirmation that great white shark attacks on humans are exceptionally rare. The incidence of great white shark attacks is very low, taking into account the wide distribution of great white sharks and the high number of people who swim and dive in the seas of the world. Great white sharks are large and powerful enough to inflict serious wounds or to kill a human. But, most of the time when they encounter a human in the water, they show no interest in attacking. Many cases have been reported of great white sharks approaching divers and bathers closely without showing any aggressive behavior.

In most cases, the attack ends after the initial contact, and the great white shark does not eat or kill the victim. The human–great white shark attack fatality rate is relatively low, between 26.1 and 33.33 percent (Burgess and Callahan; De Maddalena, 2009). Most attacks result in significant blood loss, not massive consumption by the shark. When the victim dies, it is usually as a result of shock, blood loss or other injury. Human beings are not a usual part of any shark diet, great white shark included. We presume that great white sharks do not regard humans as food, and that most attacks are not motivated by hunger.

The real reasons behind these attack behaviors remain largely unknown. Unprovoked attacks on humans by great white sharks may be attributed to various causes, including a form of play or hunting practice, perceived threat by the human, defense, interference with feeding, determination of the suitability of the victim as potential food, and mistaking the human as usual prey (Collier, 2003; Cousteau and Richards, 1992; De Maddalena, 2008; Martin, 2003). Tricas and McCosker (1984) have suggested that great white sharks mistake surfboard silhouettes for pinnipeds. Burgess and Callahan (1996) reported that a high percentage of victims wear black gear or clothing similar to the dark coloration of many marine mammals. Strong (1996b) has observed that these sharks prefer a sealshaped target when presented simultaneously with a square target. However, Collier *et al.* (1996) demonstrated that great white sharks also attack inanimate objects of a variety of shapes, sizes and colors, none resembling the shape, size or color of a pinniped. Consequently, they have suggested that these predators are in such cases determining the suitability of the prey as food. De Maddalena (2009) pointed out that great white shark attacks on humans occur even in geographical areas where pinnipeds are not among great white shark prey items. Great white sharks often bite animals that are not consumed, on many occasions killing them in the process. This happens with humans as well as with sea otters *Enhydra lutris*, northern fur seals *Callorhinus ursinus* and African penguins *Spheniscus demersus* (see page 59, "Predatory Tactics"). Martin (2003) and Collier (2003) suggested that many great white shark attacks may be a form of play or hunting practice. The opinion of the first author is that most great white shark attacks on humans are exploration and hunting practice, which are the most plausible explanations proposed to date.

Many sharks usually circle and bump the person before executing the attack. The great white shark attacks by surprise: the victim typically never sees the shark until it is too late and is overwhelmed by the unexpected assault and the violent force with which it is executed. Usually the shark approaches from below, the side or behind its victim. Frontal attacks and attacks from above are rare (Burgess and Callahan, 1996; Levine, 1996; Miller and Collier, 1980) (see page 59, "Predatory Tactics").

Shark attacks on humans occur in shallow, deep, warm and cold waters, but more attacks occur where the weather is favorable for recreational swimming or surfing. Great white sharks occasionally come close inshore near populous locations, and are encountered more often in zones where the bottom drops off very rapidly. Islands, straits, channels and shoals are also likely attack sites. The large predators swim in these areas because their prey also congregates there (De Maddalena, 2009).

Meshing is the most effective method for protecting beaches from dangerous sharks. Nets are placed parallel to the shore, and sharks are captured as they try to pass through them. Shark nets protect many beaches of Australia and South Africa. However, this method is difficult and extremely expensive to establish and maintain, since nets must be patrolled, cleaned and repaired often (in South Africa this work is done by a specific organization entitled Natal Sharks Board). Consequently, only relatively few restricted

La Jolla Cove, La Jolla, California. A 600–700 cm great white shark was responsible for a fatal attack on a diver on June 14, 1959, in these waters. The great white shark is responsible for the highest number of shark attacks on humans. Great white sharks occasionally come close to shore near populous locations, and are encountered more often in zones where the bottom drops off very rapidly (photograph courtesy Walter Heim).

areas can be protected. Moreover, there is concern about the effects of meshing on the marine ecosystem. Many alternatives to meshing have been tested, including chemical and electric repellants, and protective clothing, such as the shark-proof suit of chain mail called Neptunic, which provided inadequate or doubtful protection from great white sharks. Researchers and recreational shark divers often use cages to study the behavior of great white sharks and to photograph them. The cages are constructed of aluminum or steel, are securely attached to the boat, are usually located at the surface or at a depth of a few meters, and have viewing ports that allow photographing and filming (De Maddalena, 2008).

We know that some shark behavior involves a communication function, and these fish sometimes attempt to communicate with humans using particular signals before executing an attack (see page 66, "Competition"). These particular behaviors may function as a means of defending a great white shark's individual territory or food. Sometimes even the great white shark shows a threat display with jaws slightly open, pectoral fins depressed and back arched. Other shark species circle the prey before attacking. Unfortunately, the attacking great white shark most often does not exhibit any warning behavior and is not seen before the attack; therefore, incident prevention through interpretation of white shark behavior is almost impossible. However, we cannot forget that shark attack is a rare consequence of entering the sea, and is a risk that each diver, surfer or bather must consider.

Fisheries

Many sharks are fished commercially, and are actually overfished in many areas. As bony fish fisheries have been depleted, fishermen have compensated by increasing shark captures (De Maddalena, 2008). Among the species of sharks that are most heavily exploited are the closest relatives of the great white shark, the porbeagle *Lamna nasus*, and the shortfin mako *Isurus oxyrinchus*. The great white shark is also exploited. Many great white sharks are caught by the high seas and coastal fisheries throughout the oceans of the world. The tendency for great white sharks to investigate fishing activities and to scavenge from fishing gear makes them very vulnerable to being captured (Compagno, 2001). However, the importance of the great white shark as a fisheries species is limited because of its low abundance throughout its range. In fact, its scarcity restricts targeted commercial fisheries from intentionally seeking them for conventional bulk fisheries products, such as meat or liver oil. Unfortunately, the high value of its jaws, teeth and fins, and the good quality of its meat makes it a viable target of small-scale commercial fisheries, as well as an added value as bycatch (Compagno, 2001; De Maddalena, 2009).

The great white shark is caught and killed with pelagic and bottom longlines, specialized heavy line gear, rod-and-reel, fixed bottom gill nets, floating inshore gill nets, pelagic gill nets, fish traps (especially tuna traps), herring weirs, trammel nets, harpoons, bottom and pelagic trawls, purse seines, guns and loops (Compagno, 2001; De Maddalena, 2009).

The great white shark is or has been targeted by small-scale fisheries in several countries. An early fishery was in the region of the Eastern Adriatic Sea between the second half of the 19th century and the early 20th century at the time of the Austro-Hungarian Empire. Between the years 1872 and 1905, because of the threat that great white sharks posed to humans, the Imperial Maritime Austrian Government issued three circulars offering a reward of up to 500 florins for every great white shark captured. These circulars also mentioned other shark species, but primarily referred to *Carcharodon carcharias*. To obtain the monetary reward, the fishermen had to present their captured specimens to the Natural History Museum of Trieste to verify the species identification (De Maddalena, 2000a; 2002). Another early example was a specific fishery in the Messina Strait, Italy, in the 19th century and early 20th century, where the swordfish fishing vessels also targeted great white sharks for their meat, capturing them with a

A great white shark caught off Woods Hole, Massachusetts (photograph by Paul Galtsoff, courtesy Northeast Fisheries Science Center Photo Archive).

specific kind of harpoon (Gamberini, 1917). A third example is the small-scale fishery for great white sharks in Madagascar, from the late 20th century to the present. In this case, the sharks were specifically caught for their jaws and teeth, which were sold to tourists and European dealers (Zuffa *et al.*, 2002). The great white shark is also subject to targeted sport fisheries for game-fishing records (Compagno, 2001). Recently, the demand for shark fins has increased dramatically. Shark fins are high-priced, and this has led to the practice of finning sharks at sea, where the fins are sliced off while the rest of the body is discarded overboard. Almost all large and medium-sized sharks are fished for their fins (Watts, 2001). Great white shark fins are highly valued on the market, so this species has to be listed among the sharks that are subject to this cruel practice.

An estimated 50 percent of the world shark catch is believed to be taken accidentally while fishing for other species such as tuna (family Scombridae) and swordfish *Xiphias gladius* (Linnaeus, 1758). This unplanned capture of marine animals is called "bycatch." The great white shark is taken as a bycatch in many fisheries.

In the past, significant bycatches of great white sharks occurred in the Mediterranean Sea. The interactions of the great white shark with migrating northern bluefin tuna *Thunnus thynnus* (Linnaeus, 1758) and their consequent capture in the tuna traps in this area has been recently investigated in detail (De Maddalena, 2009). Once tuna traps were numerous along the coasts of Italy, France, Spain, Croatia, Turkey, Libya, Tunisia, Malta and Morocco (Ravazza, 2005). Captures of great white sharks in these tuna traps were relatively frequent. When a great white shark entered a trap, it was seen as a disaster by the fishermen, because the large shark could damage the nets, keep the tuna away from the trap or even scare the tuna, inducing them to break the nets. To prevent this disaster at a tuna trap in Scopello, Sicily, Italy, a cow shoulder was placed at the trap's entry, so that a great white shark could satisfy its hunger and move away (De Maddalena, 2009). Today, tuna traps have almost completely disappeared from the Mediterranean Sea due to pollution, maritime traffic and unregulated fisheries, which has resulted in these tunas changing their migratory routes and staying farther from the coasts (Ravazza, 2005). Consequently, captures of great white sharks in tuna traps have also declined (De Maddalena, 2009).

The great white shark, like many other sharks, is also taken as bycatch in longline fisheries. Pelagic longlines are single stranded fishing lines 18 to 72 km long, with an average of 1,500 baited hooks. Pelagic longlines are the most widespread fishing gear used in the open ocean and are widely used in many parts of the world to catch tuna and swordfish. In some areas, the number of sharks caught by longliners reaches 90 percent of the total capture (De Maddalena, 2008). Unlike other lamnids, the great white shark is seldom recorded from pelagic longline catches. This may be a function

A young great white shark caught off Woods Hole, Massachusetts. Many great white sharks are taken accidentally while fishing for other species such as tuna (family Scombridae) and swordfish *Xiphias gladius*. This unplanned capture of marine animals is called "bycatch" (photograph courtesy Northeast Fisheries Science Center Photo Archive).

of white shark rarity in the epipelagic zone and gear selectivity, with larger animals breaking off the longline gear. Thus, these species have seldom been caught or reported caught in the past (Compagno, 2001).

Nobody knows how many sharks are caught in the world, but the number is estimated to be enormous. Clarke *et al.* (2006) presented an estimate of 26 to 73 million sharks traded annually worldwide for their fins. Annual landings of cartilaginous fish reported to the Food and Agriculture Organization (FAO) of the United Nations amount to around 800,000 tons (Vannuccini, 1999), but the actual total is probably much higher since great amounts of catch are not recorded. This estimate does not include many thousands of sharks that are killed by fishermen annually and thrown back into the sea because in many countries numerous shark species are considered non-marketable. Industrial fishing vessels often operate in flagrant violation of fishing regulations. Moreover, many species are caught by recreational anglers. Official landing statistics of great white sharks are scarce and it is difficult to assemble a complete picture of the world great white shark catch from landing data. Compagno (1984) noted that off California, 10 to 20 or even more white sharks were killed each year as a bycatch of various fisheries. These figures span a period of about three decades up to the early 1980s. However, protection of the white shark in the 1990s may have reduced the catch since then (Compagno, 2001).

In some areas, like Mexico, wild bluefin tuna are captured and farmed in cages. Occasionally, great white sharks penetrate the cages and become trapped in tuna tow cages and in inshore tuna farm cages. In some cases, the sharks were found already dead in tuna cages. In other cases, they were deliberately killed, and in still other cases, the tuna farm staff attempted to release the large trapped predators. But it is only recently that successful releases have occurred (Galaz and De Maddalena, 2004). There are several reports of great white sharks being trapped in tuna cages. Exact numbers are not known because captures are not always reported. According to Malcolm *et al.* (2001) in Australia there are unsubstantiated reports of up to 10 to 20 captures of great white sharks by the tuna farm industry and multiple interactions each year. Further work is required to accurately estimate the number of sharks that may be trapped in tuna cages.

Utilization

Erroneously, Roedel and Ripley (1950) defined the great white shark as having no commercial value. The meat of the white shark is and has been utilized fresh, fresh-frozen, dried-salted, and smoked for human consumption (Compagno, 2001). The meat of this large predator has also been used for livestock feed, as fishmeal. Great white shark meat has often been considered of bad quality (Tortonese, 1956). Actually, the meat of this large predator is of excellent quality (Cigala Fulgosi, pers. comm. 2003), to the point that Coles (1919) described it as "*the very finest shark, or, in fact, fish of any kind that I have ever eaten, its flavor being quite similar to a big, fat, white shad*" and Wood (1959) stated to have *"never eaten better fish."* Cousins to the great white shark, the shortfin mako *Isurus oxyrinchus* and porbeagle shark *Lamna nasus*, are considered excellent to eat and command a high market price. However, great white shark meat has extremely high mercury content (Compagno, 2001). Toxic chemicals that can be absorbed or ingested by animals are passed up the food chain through consumption. Consequently, top predators like great white sharks have higher concentrations of toxins that accumulate in each organism along the food chain. This effect is further increased due to the longevity of the great white, since a maximum age estimate of 53 years has been calculated (see page 46, "Age") (De Maddalena, 2009).

The flesh has been used in traditional medicine in South Africa. White shark meat has been sold as smooth-hound (genus *Mustelus*) and swordfish *Xiphias gladius* in European Mediterranean countries, and as "shark" in California (Compagno, 2001; De Maddalena and Baensch, 2008). Because of the stigma associated with white sharks, the meat has rarely been marketed as "white shark meat," although this happened in California in the 1980s and in Sicily, Italy. The liver of the white shark has been extracted for vitamin oil, because of its high content of vitamin A. The liver also contains squalene, which, like squalene from other sharks, may be used for cosmetics, but there is no evidence that this actually happens for this species.

Great white shark fins are used in Chinese cooking to prepare a famous shark fin soup. According to Kreuzer and Ahmed (1978), fins from almost all sharks over 1.5 m in length are commercially valuable. Even though there are some species with fins that are considered excellent, preferences for fins of particular species can change from one country or one person to another. The fins of the same species can be highly appreciated by some people and refused by others. There are fins which are popular due to their high percentage yield of fin needles and their needle size, texture and appearance (Vannuccini, 1999). In a list of the preferred species for fins in major markets, Vannuccini (1999) divided them into first, second and third choices, listing the fins of the great white shark among those considered of second choice. However, this author points out that this classification must not be considered a rule but only a tendency. Great white shark fins are boosted in value because of their size and the notoriety of the species. A fin set from a large white shark may be valued at over US$ 1,000 (Compagno, 2001).

Whole jaws are used as trophies for decoration, and single teeth are used as souvenirs and as pendants. Dried great white shark jaws and teeth may reach high prices on the market, with the greatest value for the jaws and teeth of large sharks over 5 m long. In South Africa, offers of US$ 20,000 to $50,000 have been made for white shark jaws, and US$ 600 to $800 for individual teeth (Compagno, 2001).

White shark cartilage in powder form is sold as a medicinal, being promoted as a source of angiogenesis inhibitors for the treatment of cancer, even though there is no solid evidence that this product actually has this kind of property. In the past, the skin of the white shark has been utilized for leather (Compagno, 2001). Great white sharks

1—Biology, Ethology and Ecology of the Great White Shark

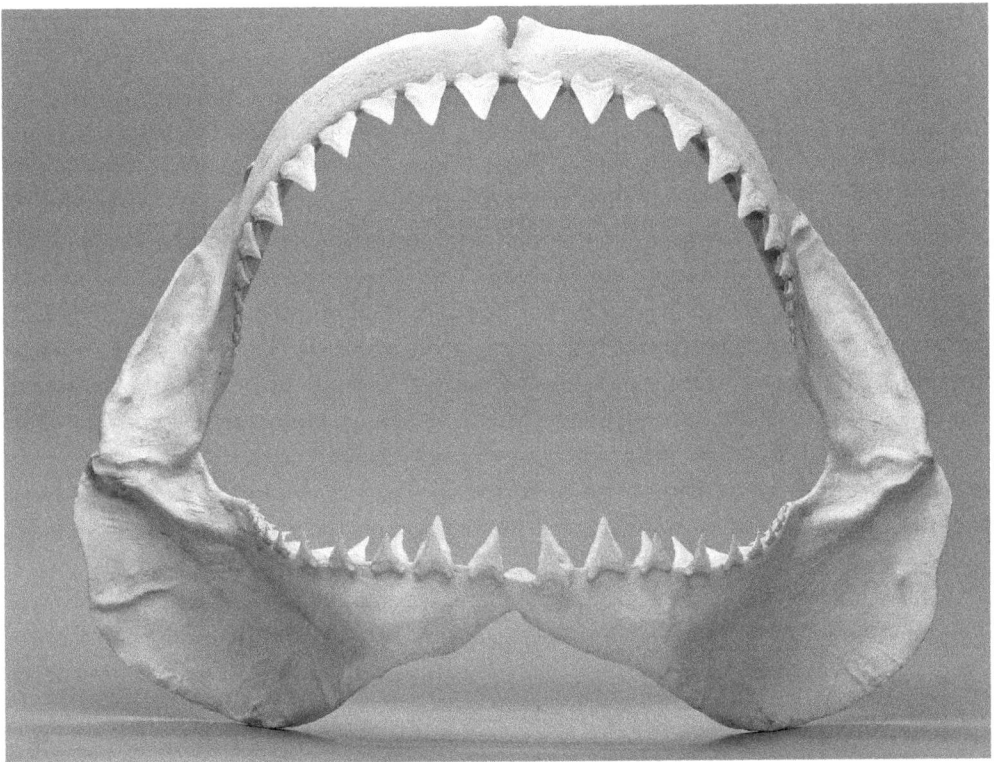

Dried great white shark jaws and teeth may reach high prices on the market. Luckily, the market of great white shark replica jaws seems to be increasing. Perhaps the most perfectly recreated are those made by Bone Clones (photograph courtesy Bone Clones, Canoga Park, California).

and their parts have been preserved, cast and put on display in museums and aquariums, as will be explored in detail.

The utilization of live great white sharks must also be considered. From the late 20th century, the great white shark has been the subject of commercial underwater cage-diving operations in South Australia, South Africa, California and Mexico. The operators that lead these expeditions allow sport scuba and snorkel divers, photographers, professional film crews and marine biologists to observe, study, photograph and film great white sharks. Regulation of ecotouristic access to great white sharks is important both for the potential danger posed by any interaction with these large predators, and for the delicate status of this threatened species. Therefore, the matter has been and still is an object of heated debate involving operators, divers, researchers and the media.

Many attempts have been made to keep live great white sharks on exhibit in aquariums and oceanariums, with poor results, which will be explored in detail in Chapter 4.

Abundance

Many species of shark have become uncommon or rare as a consequence of overfishing either the shark or its prey. In addition, humans also have a less direct, but just as harmful, effect on sharks due to depletion of resources, environmental pollution and habitat destruction. Based on their limited numbers, sharks are more vulnerable to overfishing than are bony fish. Even though few species prey on them, sharks are highly vulnerable to over-exploitation, especially since they have long sexual maturation times, low fecundity, long gestation periods, and produce small numbers of young (De Maddalena, 2008). This is especially true for the great white shark (see page 44, "Reproduction"). With heavy fishing pressures on pelagic elasmobranchs around the world, there is concern over the status of great white shark stocks. Unfortunately the total great white shark catch by ocean is not known (see page 74, "Fisheries"). Appropriate information is lacking for an assessment of this kind. Further research is needed to assess the true catch levels in each fishery and the impacts on the great white shark populations.

The size of great white shark populations worldwide remains largely unknown. The movements of the great white sharks are not well known, including the rates of emigration from one area to another, and the paths from feeding areas to mating areas.

In South Africa, the average size of the great white shark population for the region from Richards Bay in KwaZulu-Natal to Struis Bay in Western Cape was estimated at 1,279 individuals (photograph courtesy Gaspare Schillaci).

Also, there is no way to determine the rates of natural or fishing mortality (Cailliet, 1996). There is a need for a standardized means of assessing great white shark populations.

There have been some efforts to estimate the great white shark populations in the main centers of abundance in various parts of the world. The great white shark population at Dangerous Reef, the area of greatest known white shark abundance in South Australia, is relatively small, with estimates ranging from 18 to 192 individuals (Strong *et al.*, 1996). The estimates of the number of great white sharks visiting the South Farallon Islands, one of the areas of greatest white shark abundance in the Eastern North Pacific Ocean, ranged from 9 to 14 per year (Klimley and Anderson, 1996). In South Africa, perhaps the country with the greatest white shark abundance in the world, 255 individuals were recorded in the area encompassing Dyer Island and False, Struis, and Mossel bays over a 4-year period (Ferreira and Ferreira, 1996). From another study in South Africa, the average size of the great white shark population for the region from Richards Bay in KwaZulu-Natal to Struis Bay in Western Cape was estimated at 1,279 individuals (Cliff *et al.*, 1996).

Recently, an analysis of sighting and capture data of great white sharks from the Mediterranean Sea was presented (De Maddalena, 2009). This data set is the largest available for this region, containing 549 records of great white sharks. De Maddalena (2009) estimated that this population of white sharks has declined 44.74 percent in the past 10 to 20 years. In fact, there were 76 records during the 1989–1998 decade, dropping to 42 records during the 1999–2008 decade. Even taking into account that this estimate is an approximation, and that the decline of the recent years may be slightly lower than reported, there is consensus among both researchers and fishermen that the great white shark has become exceedingly rare in the entire Mediterranean Sea, and a strong decline relative to its former abundance is reported from almost all the areas where the species is distributed.

Conservation

Many nations worldwide, including the United States, have enacted laws to protect great white sharks from directed fisheries, and have prohibited the cruel and devastating practice of shark finning. Despite these laws being enacted, sadly great white sharks continue to be exploited in oceans worldwide. Even though the image of the great white shark is being used by some conservation groups in their campaign to raise funds and generate widespread advertisement for their own benefit in the worldwide media, the species is still far from salvation. Instead, it is heading toward a gradual destruction, and the idea of its total extinction seems quite possible today. The greatest threat to this species is bycatch (see page 74, "Fisheries"). Moreover, the high value of great white shark products encourages poaching, clandestine trade, and the disregard of protective laws (Compagno, 2001). By purchasing white shark teeth and jaws that most often come from illegal fisheries or in some cases were stolen from a natural history museum, collectors are partially responsible for the decline of the great white shark. For example, in the late 20th century, a small-scale fishery for great white sharks existed in Madagascar. The sharks were caught specifically for their jaws and teeth that were purchased by European dealers. Often, the fishermen did not eat the white shark meat because it was prohibited by their religious beliefs (Zuffa *et al.*, 2002).

The great white shark is gradually becoming protected worldwide. In 1991 South Africa announced protection of great white sharks within its exclusive economic zone (EEZ). The legislation made it illegal to catch or kill any great white shark, or to sell or offer for sale any whole white shark, body part or product of white sharks. In 1993, Namibia announced protection for great white sharks by banning targeted great white shark fisheries. In 1994 California introduced a temporary law that prohibited the catch

If we do not find a way to protect the great white shark, in a few decades this species may follow the destiny of its prehistoric relative, the megatooth shark *Carcharodon megalodon*. The photo shows a 1,036 cm long megatooth shark *Carcharodon megalodon* model prepared for the exhibition "Fossil Mysteries" at the San Diego Natural History Museum, San Diego, California (photograph courtesy Lollo Enstad, San Diego Natural History Museum).

of great white sharks in state waters, and the landing in California ports of white sharks caught outside state waters (Monterey Bay Aquarium, 2008b). In 1996, the great white shark was listed on Annex II (Endangered or Threatened species) of the Protocol concerning Specially Protected Areas and Biological Diversity in the Mediterranean of the Barcelona Convention for the Protection of the Marine Environment and the Coastal Region of the Mediterranean. This should result in full legal protection for the species when the Convention is ratified (Wildlife Conservation Society, 2004). But only two Mediterranean countries (Malta and Italy, see below) have implemented this listing by providing national protection for the species (De Maddalena, 2009). In 1997, California adopted permanent legislation to fully protect great white sharks in state waters, and outlawed all directed efforts to attract white sharks. Great white sharks may be collected under a permit issued for educational and scientific purposes, and accidental catch in gill net fisheries was also allowed, as it is deemed minimal. In 1997, the U.S. National Marine Fisheries Service outlawed all directed fisheries for great white sharks in East Coast waters, including Florida and the Gulf of Mexico. Great white sharks caught accidentally must be released with a minimum of injury and without taking the animal out of the water. Possession of white sharks is prohibited. Also in 1997, the Australian Government protected great white sharks throughout its Commonwealth waters. In 1998 Brazil gave great white sharks the status of endangered species (Monterey Bay Aquarium, 2008b). In 1999 Italy declared the great white shark a protected species in its territorial waters, although fishermen who accidentally catch and kill white sharks will not be prosecuted. In the same year, Malta declared the great white shark a protected species in its territorial waters. In 2000, the World Conservation Union (IUCN) added great white sharks to the Red List of Threatened Species as "vulnerable to extinction," but noted that "a global status of Endangered may be proven accurate for this shark as further data is collated." In 2000, the U.S. enacted a Shark Finning Prohibition Act to ban any person under U.S. jurisdiction from engaging in shark finning, possessing shark fins aboard a fishing vessel without the shark carcass, and landing shark fins without the shark carcass (Monterey Bay Aquarium, 2008b). In 2002 the great white shark was listed on Annex II (animal species requiring strict protection) of the Bern Convention on the Conservation of European Wildlife and Natural Habitats, meaning that it should, in due course, be included under the European Habitats Directive (Wildlife Conservation Society, 2004). In 2002, during the Convention on Migratory Species, participants demanded better international protection for the great white shark. Close to 70 nations agreed that the great white shark should be listed in the Convention on International Trade in Endangered Species of Wild Fauna and Flora (CITES) Appendix I and II to control trade in white shark parts. Additionally, all countries with white shark populations should take legal measures to prevent poaching and prevent directed or accidental catch. In 2004, white sharks were listed in Appendix II of CITES (Monterey Bay Aquarium, 2008b). In 2006, European Community vessels were prohibited from fishing, retaining onboard, transferring between vessels and landing any

The great white shark is gradually becoming protected worldwide (photograph courtesy Gaspare Schillaci).

great white shark in all Community and non–Community waters. The prohibition was also applied to all non–Community vessels in Community waters (Council of the European Union, 2006). In 2007, Mexico introduced a shark finning ban and extended protection for the great white shark. Also in 2007, New Zealand announced that great white sharks will be fully protected within its exclusive economic zone (EEZ), and that fishing by New Zealand vessels outside its EEZ will be prohibited. These regulations made it illegal to hunt, kill or harm a great white shark, or to possess or engage in the trading of any part of a white shark. However, fishermen who accidentally catch and kill white sharks will not be prosecuted (Monterey Bay Aquarium, 2008b). Again in 2007, Mauritania prohibited fishing for the great white shark using pole-and-line and surface long-line gear (European Commission, 2007).

2

Preserving and Reconstructing Great White Sharks

The Study of Great White Sharks Preserved in U.S. Museums

In 1996, in order to collect and analyze available information about the great white sharks inhabiting the Mediterranean Sea, the first author started research on the records of this large predator in these waters. This data bank was initially developed as his thesis for the completion of his studies in Natural Sciences at the University of Milan (De Maddalena, 1997). However, the large amount of data obtained from this research along with the interest in this project shown by many shark researchers, the general public and the media, convinced the first author to continue his work on this project, then named the Italian Great White Shark Data Bank. Today, this data bank includes information on 549 records of great white sharks from the entire Mediterranean Sea, representing the most complete and comprehensive study ever performed on the great white sharks in that area (De Maddalena, 2009). The data collected since the creation of the Italian Great White Shark Data Bank includes a substantial amount of information on size, distribution, habitat, behavior, reproduction, diet, fishery and attacks on humans, that have been presented in a number of scientific articles and books (Celona *et al.*, 2001; De Maddalena, 1998, 2000a, 2000b, 2002, 2006a, 2006b, 2009; De Maddalena *et al.*, 2001, 2003; Galaz and De Maddalena, 2004; De Maddalena and Révelart, 2008; De Maddalena and Zuffa, 2008).

The location and the study of white sharks preserved in natural history museums has always been a fundamental part of this research program. Many European museums own at least some great white shark items in their collections, and some of these institutions hold remarkable specimens. A detailed report on 109 great white shark specimens preserved in 49 European institutions has been presented in De Maddalena (2006b, 2007). After completion of this work, the first author has decided to continue researching great white shark materials preserved in museums outside of Europe, with the focus of his study in the United States, the nation that has the highest amount of preserved items from this large marine predator in the world.

In 1939, renowned shark specialist Stewart Springer wrote about the great white shark: "It is unlikely that many specimens will be preserved, and unless fairly complete and accurate data are secured from those that do happen to fall into the hands of ichthyologists, it is probable that our knowledge of the most formidable of existing sharks

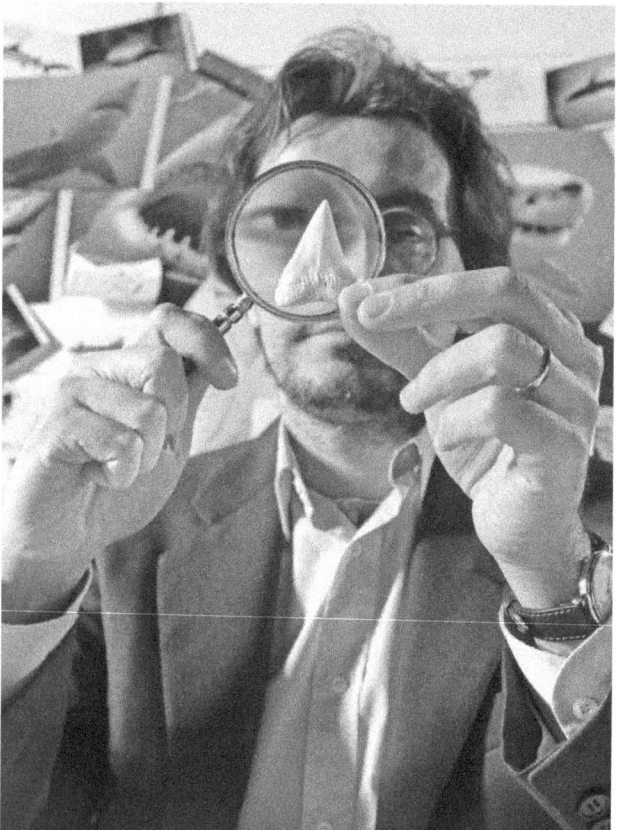

Alessandro De Maddalena examines a tooth of a great white shark. The study of white sharks preserved in museums has always been a fundamental part of the work done in assembling information for the Italian Great White Shark Data Bank. After completion of his work on the materials of white sharks in European museums, the first author has decided to continue researching great white shark materials preserved in museums outside of Europe, with the focus of his study in the U.S.A. (photograph courtesy Nicola Allegri).

will remain in its present unsatisfactory state" (Springer, 1939). Taking into account that the abundance of this species has decreased since the time when Springer made this statement, the authors felt that a complete catalogue of the great white shark materials preserved in the United States of America was needed.

Therefore, a total of 366 institutions have been contacted, including all museums, institutes of Natural History, Zoology, Biology, Anatomy, and aquariums and oceanariums across the entire United States (see Appendix I, "Institutions Contacted"). Most contacted institutions have been very responsive, offering their collaboration for the completion of this project. Material held by private citizens has not been considered in this study. It was decided to restrict the focus of work to the material that has been preserved in museums and scientific institutes and is therefore available to be studied by researchers. All available information on this material is presented in this book.

The data for this book was obtained primarily from institution catalogues and from their staff, collectors and determiners of great white shark materials. Information was also obtained from scientific publications and any other available source. For each item, when possible, the following data was collected: institution name, institution address, storage or exhibit location, including catalogue number, type of material (whole specimen, set of jaws, teeth, vertebrae or other anatomical parts), preservation method (liquid, dry, taxidermied, model reproduced from a mold of a real specimen), date of capture, place of capture (with exact position, latitude and longitude, when available), depth of sea, depth of capture, sex, total length (TL), weight, stomach contents, sources ("pers. comm." refers to a personal communication received by the first author in the specified year), date of acquisition, collector, determiner, and any available photographs. Measurements on the jaws and teeth are reported just in a few cases, and include enamel height of the largest upper tooth (UAE1 and UAE2) and dried upper jaw perimeter (DUJP), using the methods of Mollet *et al.* (1996).

All requested measurements were given in centimeters. In addition, when the original source of the data reported the shark length in feet, the original length in feet is reported in brackets. The weight is expressed in kilograms. When the original source of the data reported the shark weight in pounds, the original weight in pounds was reported in brackets.

The data was requested per a form that is shown in Appendix II.

Basic Information on Great White Sharks Preserved in U.S. Museums

In conclusion, 160 white shark items from 26 institutions in 14 states have been found. The states where the great white shark items are preserved include Alaska, California, Connecticut, Florida, Illinois, Massachusetts, New York, North Carolina, Pennsylvania, South Carolina, Virginia, Washington, Washington, D.C., and Texas.

The material preserved in the museums and research institutes of the United States includes whole specimens as well as parts of specimens. These parts include heads, tails, fins, skin, dermal denticles, skeletons, chondrocrania, jaws, teeth, tooth fragments, vertebral columns, single vertebrae, eyes, gill arches, claspers, testes, hearts and samples of muscle tissue. Besides actual shark items, replicas that were made from a cast of the original specimen and photographs are included in some collections. Stomach contents from a few sharks were also part of some collections. The specimens found in some collections include great white sharks that were kept in captivity in American aquariums, and were included among those listed in Chapter 4. However, this is not indicated in the museum catalogues, but has been confirmed in a few cases.

Most of the material is preserved in excellent condition and includes numerous interesting items. Nevertheless, the amount of material that is on display is very small. Most items are not currently on exhibit to the public. There are also a small number of items listed in museum catalogues, which are currently missing. The reasons they are missing are unknown, and the possibility that they may still be present in the museum collections, perhaps stored in some unknown location, cannot be excluded. Therefore, the missing items have been included in this book as well.

It is inevitable when preparing such a work on museum materials, that the data reported by the museums is often incomplete. This can happen for different reasons.

A life-size replica of a 548.6 cm pregnant great white shark being installed by volunteers and museum staff at the Humboldt State University Natural History Museum in Arcata, California. This replica is on loan from the California Academy of Sciences (photograph courtesy Debbie Paselk, Humboldt State University Natural History Museum, Arcata, California and California Academy of Sciences, San Francisco).

Often the items were acquired by a museum with none or little information about them. In other cases, the logging of an item into a museum catalogue with little information was left to the discretion of the museum staff. So, some items did not have catalogue numbers. In other cases, the data was supposed to be available at present, but the museum staff was unable to provide the information we requested even though the request was made long ago, mainly for lack of personnel.

In a few cases, other shark items were wrongly identified as great white sharks, such as in the case of two jaws preserved in the Museum of Comparative Zoology at the Harvard University in Cambridge, Massachusetts, with catalogue no. MCZ 153577 and MCZ 153578, that we re-identified as belonging to bull sharks *Carcharhinus leucas*.

The great white shark items in U.S. museums were preserved using different methods, including fixation in formalin followed by permanent storage in aqueous solutions of ethyl or isopropyl alcohols, freezing, drying, taxidermying or skin-mounting, embalming, and creating a model of the shark that was cast from a mold of the original specimen. Moreover, models not produced from original specimens, but sculpted by artists, are in some cases on exhibit for educational purposes. Details of the methods of preservation or preparation of great white shark items are explored in the following chapters.

Preserving Great White Sharks in Liquid

The main problem with preserving whole great white sharks is, without a doubt, their large size. In the case of newborn and juvenile specimens, their smaller size allows whole preservation. Larger great white sharks present an enormous storage problem, and it is usually impossible to preserve them intact. However, parts of the shark, such as skin, heads, fins and vertebral columns of even large sharks can be accommodated in containers without much problem. Compagno (1984) suggested a procedure for preparing a large shark for compact storage. The viscera and most of the muscle mass from the pectoral fin bases to the second dorsal and anal fin should be removed with a filleting knife, leaving a long dorsal strip of skin connecting the head to the first dorsal fin, second dorsal, caudal peduncle and caudal fin, and a short ventral strip connecting the pelvic fin bases, anal fin and caudal peduncle. The vertebral column should be stripped of excess flesh and severed at the head and caudal peduncle, then cut into smaller sections if necessary (Compagno, 1984).

Preservation means maintaining the state of fixed tissues. So, the first step for preserving great white sharks is the fixation of tissues. Fixation is the process of coagulating cell contents into insoluble substances (usually by cross-linking proteins), preventing autolysis and breakdown of tissues (Fink *et al.*, 1979). By the beginning of the 20th century, fixation in water solutions of formaldehyde gas or formalin, followed by permanent storage in aqueous solutions of ethyl, isopropyl or n-propyl alcohols became the standard method of preservation (Compagno, 1984).

For best results, sharks should be fixed as soon after death as possible, although they can be frozen or even covered with ice to halt or retard putrefaction until the specimen can be fixed. Excess freezing will dehydrate unprotected specimens. Consequently, sharks to be frozen for considerable periods should be sealed in plastic bags with some water. In hot climates, it is especially important to preserve or freeze specimens quickly, as they can deteriorate in a matter of hours. Specimens should be kept cool, in the shade, and iced or covered with wet cloth or burlap if they cannot be immediately fixed or frozen (Compagno, 1984).

As it is used in ichthyology, formalin is a saturated water solution of formaldehyde with a chemical formula of CH_2O. Formalin is the standard fixative used in ichthyology, produced by dilution of formaldehyde solution with water, usually by adding nine parts water to one part formaldehyde solution. This results in a 10 percent formalin solution (Fink et al., 1979). Formalin appears to function as a fixative by forming cross-links between adjacent protein chains, denaturing them and thus deactivating them. Autolysis, which is the destruction of a cell by its own enzymes, is stopped and the proteins are coagulated, preventing breakdown of tissues (Fink et al., 1979). Formaldehyde oxidizes into formic acid, and since this process occurs more rapidly in dilute solutions, formalin should be prepared from stock only as needed (Fink et al., 1979).

Formalin is a hazardous substance and should be handled with extreme care. Long-term exposure to formalin may result in mild to very severe allergic reactions requiring medical attention (Fink et al., 1979). Therefore, formalin should be handled in a well-ventilated place using protective clothing and safety glasses (Compagno, 1984). Containers for fixation should be kept tightly closed with tight-fitting lids to prevent contact with air, both to prevent escape of toxic formalin fumes and to maintain proper strength, since formaldehyde or formalin which has oxidized to formic acid is weaker than the standard unoxidized solution (Fink et al., 1979). Large containers, such as barrels and tanks, are necessary to fix and house a whole great white shark or its parts. Formalin will quickly corrode ordinary steel. Therefore, containers should be made of stainless steel, which does not stain, corrode, or rust as easily as ordinary steel, or from steel with an acid-proof coating (Compagno, 1984).

A label for the shark should be made in pencil, with catalogue number, date and location of capture, depth of capture, total length (TOT), sex and collector name. The label should be tied to the caudal peduncle of the great white shark or placed inside its mouth or inside a gill slit (Compagno, 1984). Since specimens accompanied by a faded or disintegrated label are of little value to most ichthyologists, it is important that paper used for labels be of sufficient quality to withstand contact with specimens, handling, and exposure to formalin and alcohols. As a alternative to pencil, a good quality carbon ink for labels, compatible with the label paper, can be used (Fink et al., 1979).

2 — Preserving and Reconstructing Great White Sharks

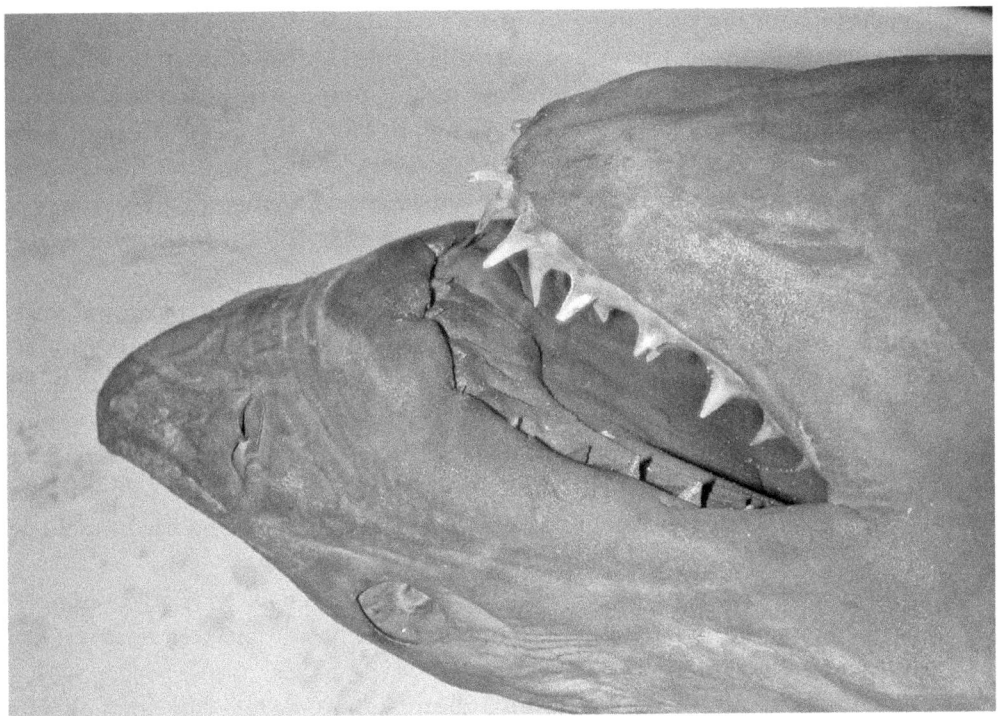

Containers should be made of stainless steel, which does not stain, corrode, or rust as easily as ordinary steel, or be made from steel with an acid-proof coating. This 203.0 cm TL great white shark, captured on April 18, 1974, off Shackleford Banks, North Carolina, was stored for many years in a steel tank that rusted and thus imparted a reddish brown coloration to this specimen. It is preserved at the North Carolina State Museum of Natural Sciences in Raleigh, North Carolina, with cat. no. NCSM 28427 (photograph by Gabriela Hogue, courtesy North Carolina State Museum of Natural Sciences).

With a syringe, a quantity of formalin is injected into the body cavity, the muscle masses of the body, tail, and fin bases, and into the head to preserve the brain (Compagno, 1984). Dilute 10 percent formalin is typically used for injection, though higher strength formalin can be used. According to Compagno (1984), 1:4, 1:2 or even undiluted concentrated formaldehyde is very effective and preferable for preventing putrefaction in hot climates. However, other researchers suggest that such high concentrations may harden outer tissues and slow the penetration of formalin to inner tissues (Fink *et al.*, 1979). If a syringe is unavailable, formalin can be introduced through several small holes or slits on the sides and body cavity of the shark made with a knife, scalpel or probe, preferably on the righthand side. The left side is generally used for illustration, although at least one small slit on the left side of the body cavity should be made (Compagno, 1984). This method is not recommended for very large specimens, because it is very difficult to distribute formalin throughout the body, since diffusion transports the formalin only 1–2 inches into the body. For large specimens, it is much better to pump formalin throughout the body using the shark's circulatory system. This can be done by cutting one of the major arteries and attaching a small hose to that artery. This hose can then be attached to a small, low-power peristaltic pump, which circulates the formalin throughout the body, just as the heart pumps blood (Charles F. Cotton, pers. comm. 2009). After the formalin injection, the shark is then positioned flat on its abdomen with fins spread in the preserving container and enough dilute formalin is added to cover it. The fins can be pinned out on pieces of Styrofoam or other soft material for positioning if necessary (Compagno, 1984).

Specimens should remain in formalin until the tissues are fully penetrated and hardened. The length of time required for saturation depends on the temperature and size of the specimens. Fixation can last for at least two weeks for small sharks less than 1.5 m to a month or more for larger specimens. A layering effect of pH in formalin solutions used for specimen preservation was noted, with formalin near the specimen at the bottom of a container at about pH 6.4, while in the upper parts of the solution, pH was 8.4. This suggests that the container in which specimens are being fixed should be stirred periodically to mix the solutions (Fink *et al.*, 1979).

For long-term preservation, chemicals less toxic than the initial fixatives are used for ease of handling the preserving fluid and to circumvent side effects of fixatives, like decalcification of bone in fishes kept in formalin.

Large specimens are usually placed into full strength storage alcohol either directly from formalin or following a wash period. Washing fixed specimens for a lengthy period of time removes a large part of the formalin and frees, to an undetermined degree, the active sites on proteolytic enzymes. Leaving a trace of formalin in alcoholic specimens may significantly lengthen their storage life. This practice, however, can cause problems for people with formalin allergies (Fink *et al.*, 1979). A short wash in water is therefore probably the best solution. The volume of the great white shark should not exceed half of the volume of the preserving fluid (Compagno, 1984).

Most institutions use aqueous solutions of ethyl or isopropyl alcohols for permanent storage of great white sharks. Ethyl alcohol, also called ethanol, has a chemical formula of CH_3CH_2OH, is one of the principle preservatives used in fish collections, and is used in aqueous solutions usually at 70–75 percent ethanol. In general, ethanol is considered more tolerable to work with than isopropanol, in terms of smell and effect on skin. Also, ethanol may not leach color as fast as isopropanol. The problem with ethanol is that it is a fire hazard. Therefore, areas where undiluted stock is stored require explosion-proof electrical fixtures and switches, and smoking should not be allowed where ethanol is being used (Fink *et al.*, 1979).

Isopropyl alcohol, also called isopropanol, has a chemical formula of $CH_3CHOHCH_3$, and is one of the principle preservatives used in fish collections. Isopropanol is used in aqueous solutions of usually 40–50 percent isopropanol and is preferred by many users because it is less expensive than ethanol. It is also less volatile than ethanol, and thus the fire hazard is reduced and costs due to loss by evaporation are less. Unlike specimens preserved in ethanol, specimens preserved in isopropanol are not suitable for most histological techniques (Fink et al., 1979).

Because of the fire hazard associated with high concentrations of alcohol or its high cost, some institutions store sharks in a weak formalin solution. If formalin is used as a long-term preservative, bottles and tanks should be kept tightly sealed, and it should be buffered with calcium carbonate. Because formaldehyde oxidizes into formic acid, and since this process occurs more rapidly in dilute solutions (like the 10 percent most frequently used), formalin should be prepared from stock only as needed and buffered if it will be used for any length of time. The acidity of unbuffered formalin will cause decalcification of hard tissues. A buffer is a chemical system that prevents change in the concentration of another chemical substance, e.g., proton donor and acceptor systems serve as buffers preventing marked changes in hydrogen ion concentration. The buffer of choice for formalin is calcium carbonate, because it is readily available and inexpensive (Fink et al., 1979). Storing great white sharks in formalin is far less desirable than storing them in aqueous solutions of alcohol because of the excessive hardening of soft tissues, brittleness of fins, and decalcification of hard tissues (Compagno, 1984).

A 1 percent solution of propylene phenoxcetol with 5 percent ethylene glycol in water has been used with some success as a substitute for formalin or alcohol storage after formalin fixation (Compagno, 1984).

Sometimes the evaporation of alcohol and its high cost force a museum staff to partially refill the container or tank where the shark is stored. There are serious problems with the practice of partially refilling containers in which some alcohol has evaporated. During the early stages of fluid loss, the alcohol portion is lost first because it is more volatile than water, thus the solution becomes dilute. In drastic cases, only water may be left in the bottle. If "topping-off" must be done, the alcohol solution to be added should be stronger than the original concentrations so that the preservation concentration is maintained. Oxygen content of storage alcohol may be a potential source of specimen degradation, particularly in loss of color due to pigment oxidation. Therefore, "topping off," and any practice that introduces air into the preservative could contribute to a shortening of specimen life (Fink et al., 1979).

One often overlooked problem in collection curation is that of the collection environment. Maintenance of proper light and temperature conditions can significantly affect specimen preservation. Light may be the most significant factor contributing to fading of specimens as specimens kept in lighted rooms tend to fade. Temperature is second in importance to alcohol strength in maintaining specimens in good condition since high temperatures speed autolysis. Temperature fluctuations can cause sealed closures to loosen, allowing alcohol evaporation. So, extreme temperature fluctuations in collection rooms need to be avoided. Ideally, specimens should be maintained in the coolest and darkest environment possible. Such conditions tend to reduce specimen fading and alcohol evaporation. Additionally, low temperatures retard the action of temperature sensitive proteolytic enzymes (Fink et al., 1979).

Specimens stored in alcohol that show signs of decomposition may be refixed. Formalin at 10–20 percent concentration is usually used and the time of refixing is size and temperature dependent. Specimens that are refixed are usually firmer and in better condition than before refixation. There seems to be no harm done to specimens and there is a good probability that refixation can prolong the usefulness of museum specimens (Fink *et al.*, 1979).

Charles F. Cotton told us how he fixed and preserved a 250 cm TOT male specimen that was captured on June 16, 2004, off Smith Island Shoals, that is still in the collection of the Virginia Institute of Marine Science in Gloucester Point, Virginia (catalogue no. 11343) (see page 162, "Virginia Institute of Marine Science, College of William and Mary, Gloucester Point, Virginia"). Charles F. Cotton used large syringes and injected formalin throughout the musculature and head. He also made a large incision along the ventral surface to allow formalin to penetrate into the internal organs. Later, Cotton used twine to sew this incision so that the organs cannot fall out of the specimen. He was particularly careful to ensure that formalin was injected into the stomach and ring valve. After a month or more in the fixative, the specimen was flushed several times with water spanning several days, and then it was stored in 70 percent ethanol (Charles F. Cotton, pers. comm. 2009).

Preparing Dried Great White Shark Jaws

Dried great white shark jaws in museum collections show various conditions of preservation. Some are perfectly cleaned while others are in very bad condition, with dirty cartilage and often with the rows of replacement teeth covered by hardened connective tissue. In order to prepare a set of great white shark jaws for the exhibit and to obtain a quality result (a perfectly cleaned set of jaws without any awful smell), it is best to start with fresh jaws taken from a fresh great white shark head. Cleaning an already dry set of jaws in bad condition is still possible but is surely more delicate and difficult work. In this section, the process used for preparing a great white shark set of jaws and the method for restoring an already dried set of jaws will be explained. The process is described by Italian shark jaw preparer Francesco Guerrazzi.

Patience and care are the fundamental requirements for this kind of work. In fact, the first step is the manual removal of the jaws from the rest of the skull and from all the tissues that cover them. For removal and processing, a scalpel equipped with several blades may work well when preparing small shark jaws. A sharp fish-processing knife and protective gloves are needed when working on a great white shark set of jaws. A scalpel may be useful as an additional tool to execute the final steps of the cleaning. Boiling the jaws must be avoided, because when heat is applied to the connective tissue (tooth bed) of the jaw it loses strength and this results in the detachment of all the teeth from the jaw. If only the teeth are to be preserved, then boiling the jaws is a good method for having all teeth detached.

The head of the shark has to be washed carefully with a stream of water to remove all the blood and mucus. This is done prior to the delicate task of removing the jaws with the musculature from the head. The upper jaw must be manually protruded, by pressing from the exterior below the shark's eyes. An incision along the jaw perimeter is then made with a sharp knife, being careful not to cut on the jaw itself, but on the exterior. The incision must start centrally on the upper jaw, cutting toward the corners on the right and left side. While cutting, two cartilaginous processes or projections of the upper jaw will be found, both on the right and left side. The incisions must be made around these processes, cutting the ligament that connects these processes to the chondrocranium. The incision is continued until it reaches the joint with the lower jaw, being careful to stay well outside of the upper jaw. This process is repeated for the lower jaw, cutting from the symphysis (which is the joint along the median line at the front of the upper and lower jaws) to the corners on each side, taking care not to cut the jaw cartilage. At the corner of the jaw, it is necessary to use meat scissors. The upper cartilages of the hyoid arch are called the right and left hyomandibulars and form a bridge attaching the jaws to the chondrocranium (see page 29, "Digestive System"). It is necessary to cut the right and left hyomandibulars in half, leaving them connected to the jaws. When this operation is complete, the jaws are freed from the head, and if the cutting has been well executed, the jaws will be intact, covered by skin and meat, with the two portions of hyomandibulars attached to them. At this point, it is important not to remove skin, muscular and connective tissues. The cleaning will be done subsequently.

The jaws are then washed carefully and thoroughly with water to remove blood and the loose pieces of tissue prior to the cleaning process that will last approximately two hours or more. To begin the cleaning of the jaws, the knife should be used to create a shallow incision along the outside of the upper jaw, where the skin and the jaw connect. With the help of a pair of tweezers, the skin, muscle and the aponeurosis, or the layer of flat broad connective tissue that covers the jaw, must be stripped away. In order to work with precision, the cutting should alternate with washing the work area with water. A scalpel may be used instead of a knife in areas that require a smaller cutting tool for delicate parts of the cleaning. This procedure is repeated for the lower jaw, removing the meat, skin and connective tissue from the outside portion of the jaw. After removal of all tissues covering the outside of the cartilage jaws, the connective tissue that protects the rows of replacement teeth on the inside of the jaws must be removed. It is important that the removal of the connective tissue protecting the rows of replacement teeth is extremely accurate and complete. This long operation may be done with the help of a pair of tweezers, raising and cutting connective tissue at its origin in the inner surface of the jaw, exposing all the teeth down to the last row, which were most recently formed. If some trace of this connective tissue remains, it will tarnish the final appearance of the jaw set and its decomposition will cause an awful smell which is hard to eliminate. It is now necessary to clean the cartilaginous processes of the upper jaw by removal of the soft white layer of external connective tissue.

The two portions of hyomandibulars attached to the jaws should now be removed. An incision should be made at the hyomandibular extremity, close to the jaw cartilage, being careful not to cut the ligament which connects the upper and lower jaws. After removal of the right and left hyomandibulars, the jaws have to be washed and rubbed energetically with a brush so that any tissue remaining is removed. At this point, the jaws should be almost clean of skin, muscle and connective tissue, with only small fragments remaining.

In order to eliminate bacteria and possible minute tissue fragments, the jaw set is washed in an aqueous bleach solution of 75 percent water and 25 percent bleach. A large container is required as the jaws are fully immersed in this solution. The jaws are allowed to soak for a time ranging from a half hour to over one hour, depending on the size of the jaws. During this first washing with bleach, the jaws are periodically removed from the solution and scrubbed to remove remaining tissue. Afterwards, the jaws are washed in cold water and again immersed in the solution. This operation must be executed until all of the tissue is completely removed. It is necessary to ensure that the solution is not too concentrated, as this may damage the cartilage. Different bleach concentrations may be used depending on the desired result. After this washing and scrubbing in the bleach solution, the jaws should be perfectly clean.

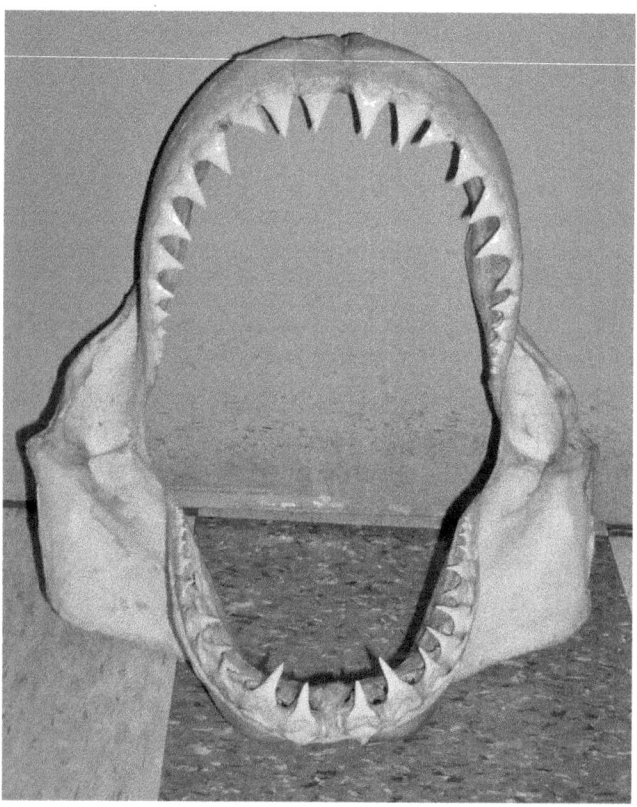

At this point, a final washing needs to be done in order to whiten the cartilage and ensure a more complete disinfection of the jaws. The jaws are then immersed in a

A set of great white shark jaws, preserved at the Natural History Museum of Los Angeles County in Los Angeles, California (photograph courtesy Antonella Preti, courtesy Natural History Museum of Los Angeles County).

12.5 percent by volume aqueous hydrogen peroxide solution. Once the jaws are immersed in the solution, the cartilaginous tissue will react by forming foam and will slowly whiten. During this step, in order to check the whitening status, it is important to alternate washing the jaws with water and immersion in the aqueous hydrogen peroxide solution. Like the previous washing with bleach, it is necessary to ensure that the hydrogen peroxide solution is not too concentrated, in order to prevent damage to the cartilage. This washing in hydrogen peroxide aqueous solution allows a complete cleaning and disinfection of the inner surfaces of the jaws, including the inside rows of replacement teeth. If the jaws smell bad upon drying, it will be necessary to repeat the washing with bleach and hydrogen peroxide aqueous solutions respectively.

After the cleaning and washing are completed, the jaws are dried. After drying the jaw set with a cloth, the jaw set should be attached to a wooden surface in the desired position. The position should be natural, and not unnaturally widened with the aim of increasing gape size by bending the jaws at the symphysis. In order to fix the jaws in position, spits may be used, forming a cross to prevent the jaws from closing. If necessary, the jaws may also be blocked at the joints with a wood wedge and fastened to the wooden surface. After fixing, it is suggested to spray its inner surface with a common anti-bacterial disinfectant. This will also repel insects during drying. Then the jaws must be covered with a transpiring cloth and be left to dry, but not in the direct sunlight, because this will cause a yellow coloration. In order not to attract insects, it is suggested to dry the jaws inside an enclosure, rather than outside in the open air. After drying, the jaws will be ready to be preserved in a museum collection, to be put on display or to be studied.

Restoring a set of jaws that has been dried without an adequate cleaning is possible, but it is important to evaluate the general status of the item prior to restoration. If the jaws do not show any damage to the cartilage and the teeth appear to be strongly attached to the jaw, the jaws need to soak in the aqueous hydrogen peroxide solution, and possibly in the aqueous bleach solution if a stronger cleaning is needed. The jaws should soak until the tissue remains to be removed are softened, then cleaned and dried as described above. When the jaws that need to be restored are in bad condition, with holes, breaks or with unstable teeth, the restoration process is extremely difficult and may damage the jaw set. In fact, it is possible that after the washing, some teeth may fall out and the cartilage may, in some places, partially fall to pieces, forcing more extensive work to restore the item (Francesco D. Guerrazzi, pers. comm. 2008, 2009).

The preparation of dried great white shark jaws varies from one person to another and from one institute to another. So, for example, a different process is followed at the Natal Sharks Board, in Umhlanga, KwaZulu-Natal, South Africa. The steps used by the Natal Sharks Board technician to prepare dried great white shark jaws can be described as follows. First, the jaws are removed from the specimen. Then, all the meat, skin and connective tissue are mechanically removed from the jaws. The jaws are rinsed with fresh water and soaked in hydrogen peroxide solution (approx. 100 ml peroxide:

20 l water) for one day. The jaws are removed and rinsed with fresh water. The jaws are positioned with wooden sticks that are placed horizontally and vertically to prevent deformation of jaws. After positioning, the jaw set is allowed to dry in sun for a minimum of 3 days. Once dry, a first coat of resin (plus catalyst) mixed with carbosil (white powder to thicken the resin) is applied. The first coat is allowed to dry for a minimum of one day (not in the sun). Sandpaper is used to smooth this first coat. A second coat (with no carbosil) is applied and smoothed with sandpaper after drying. A final coat is applied (with no carbosil) with no sanding operation. The work is now complete and the jaw set is ready for display (Sabine Wintner, pers. comm. 2009).

Taxiderming Great White Sharks

Taxidermy is a general term describing the many methods of reproducing a lifelike three-dimensional representation of an animal for permanent display. The word "taxidermy" is derived from two ancient Greek words: "taxis," which means movement, and "derma," which means skin, referring to the fact that many taxidermy procedures involve removing the natural skin from the specimen, mounting this skin over an artificial armature, and adjusting the skin until it appears lifelike. In other cases, the specimen is reproduced completely with man-made materials.

In the 18th and 19th centuries, museum great white sharks were often skinned and the skins dried and stuffed. This was supplemented and largely replaced by fixation in formalin followed by permanent storage in aqueous solutions of alcohol or by recreating the shark via cast from the original specimen. Today, skin-mounted great white sharks are rarely prepared. One example of such skin-mounted specimens (without catalogue number) was a six foot great white shark from Woods Hole, Massachusetts, that until the late 1940s, was on exhibit to the public in the New England Museum of Natural History (now the Boston Museum of Science) in Boston, Massachusetts (see page 148, "Museum of Science") (Larry Bell, pers. comm. 2008; Bigelow and Schroeder, 1948; Carolyn Kirdahy, pers. comm. 2008).

Austrian taxidermist Ernst Hofinger described the preparation of a skin-mounted great white shark (De Maddalena, 2007). Because of the size of the great white shark, the taxidermist needs a large work area and is often forced to do some work outside. A big freeze-dryer is needed to freeze-dry the shark whole, but the freeze-dryer may be too small for larger specimens. Therefore the body is dried in a conventional way. The fins are removed and freeze-dried without changing their natural form in a freeze-dryer. The shark is then skinned. The 10 mm thick skin of the white shark is thinned to approximately 3 mm thick with a knife. Then the skin is thinned once again to approximately 1 mm with a circleknife. A form is made out of polyurethane foam blocks, which are cut with a saw. When the form of the shark begins to emerge, it is then smoothed with a plane to the final shape. To make the form slippery it is buttered with wet clay.

A replica cast made from a 484 cm TL male great white shark caught east of Block Island, Connecticut, on August 5, 1983. A fiberglass epoxy resin mold was manufactured at the Mystic Marinelife Aquarium by a Long Island firm. The final taxidermy and painting was done by Peter Wilson of Taxidermy Plus in Hamden, Connecticut. The replica is on exhibit at the Department of Marine Sciences of the University of Connecticut in Groton, Connecticut, without cat. no. (photograph courtesy J. Evan Ward, Department of Marine Sciences of the University of Connecticut, Groton).

The tanned skin is pulled over and fixed on the shark form. Next, the real teeth are installed on the form. The shark, without fins, is dried in a storeroom. After drying, any skin injuries are concealed and the glass eyes are installed. The mount is now ready to be painted with the base colors. The freeze-dried fins are installed later and the shark is painted with an airbrush to the final color (De Maddalena, 2007).

Another excellent option to preserving the original specimen is to prepare a great white shark replica that is reproduced completely with man-made materials, by recreating the shark from a cast of the original specimen. In this way, all proportions are exactly reproduced, which makes this kind of model the best possible option to meet both scientific and educational purposes (De Maddalena, 2007).

For over fifty years the master taxidermists at Lyons and O'Haver, Inc., located in San Diego, California, have been mounting sharks. In the 1950s and 1960s, they were selected by the Bureau of Fisheries and the United States Navy to create replicas of sharks and porpoises for water tunnel testing. By creating a plaster mold from the

For over fifty years the master taxidermists at Lyons and O'Haver, Inc., located in San Diego, California, have been mounting sharks. Perhaps the most impressive white shark mount was a bust mount from a 518.2 m white shark, extending from the tip of the snout to just behind the pectoral fin on one side. They were commissioned to recreate the original specimen using a bust mount cast and a blank for the rest of the shark body. The final product was fitted with a real jaw set prior to crating and shipping overseas to a Japanese museum (photograph courtesy Lyons and O'Haver, Inc., La Mesa, California).

original specimen, a plastic cast is made. For larger fish, a fiberglass mold is used. The off centerline fins, the pectorals and pelvics, are removed prior to molding the body and molded separately. Once cast, these fins are reattached. One aspect of the Lyons and O'Haver mount is the detail in the mouth as they mold the interior of the mouth capturing details such as gill arches.

Sharks are different from other fish in that the cartilaginous body is soft and flexible. When cast, they tend to collapse inward. This usually requires the addition of filler to bring the shark cast back to the original shape, based on reference photos. If multiple copies of the shark are to be made, a second mold is made from the first cast that has been reshaped. The original jaw set is included with the mount if a single copy is made. Otherwise, the teeth are cast separately and bonded to the body cast later. Once assembled, the cast is masterfully airbrushed to the original color.

Over the years, Lyons and O'Haver have mounted many white sharks, mostly smaller fish. One larger specimen was a 365.8 cm (12 ft), which hangs in the lobby of the Birch Aquarium at the Scripps Institution of Oceanography. Besides white sharks,

Some steps in the preparation of the exhibition, "Megalodon: Largest Shark That Ever Lived," produced by the Florida Museum of Natural History, University of Florida, in Gainesville, Florida, with support from the National Science Foundation. Figure 1—Sculptor Jeff Huber adds some touch-up paint to the belly of a nearly finished great white shark replica (Florida Museum photograph by Mary Warrick). Figure 2—The finished great white shark model (Florida Museum photograph by Eric Zamora). *Following page:* Figure 3—The great white shark model positioned in the exhibition. In the background, some reconstructed jaws of prehistoric *Carcharodon* (Florida Museum photograph by Eric Zamora).

Fig. 3

they are masters at replicating makos, blue sharks, threshers and any other shark, given a picture and a specimen.

Perhaps the most impressive white shark mount was a bust mount from a 518.2 m (17 ft) white shark, extending from the tip of the snout to just behind the pectoral fin on one side. The cast alone weighed 272.2 kg. One of the mounts was displayed prominently in a local restaurant, the Fish Market. The mold is still available for more copies. They were commissioned to recreate the original specimen using a bust mount cast and a blank for the rest of the shark body that required substantial shaping. The final product was fitted with a real jaw set prior to crating and shipping overseas to a Japanese museum. Besides being master taxidermists, Lyons and O'Haver are also master sculptors. They sculpted a similar size whale shark for the same customer (Mike O'Haver, pers. comm. 2009).

Because the white shark is becoming a protected animal, specimens for mounting are becoming rare. Consequently, copies from existing molds are one of the few remaining ways of obtaining a mount. Lyons and O'Haver can create a mold from an existing mount in order to create replicas. The same is true for other protected animals, such as dolphins. Perhaps the signature to the Lyons and O'Haver business is one of their advertisements, which is a white shark cast mounted atop the company van!

Back in the mid to late 1970s, Sea World was buying large whites from commercial fishermen, and freezing them in displays for the public. It seems they did this in all three Sea World parks (San Diego in California, Aurora in Ohio, and Orlando in Florida). In 1976, a 518.2 cm (17 ft) great white shark with an entire juvenile elephant seal *Mirounga angustirostris* in its stomach was caught. The great white shark was necropsied at Sea World in San Diego, and then displayed to the public as a frozen specimen. Sea World's taxidermist Greig Hill did a wonderful job of presenting this 3175.1 kg (7,000 lb) fish in a natural swimming position. After a year, this specimen was quite freezer-burned and in pretty bad shape. The jaws were removed and the body disposed of. The staff at Sea World also made molds from some of the sharks for making life size models in the future. Most of the great white sharks were caught offshore in Southern California around the Channel Islands, including San Clemente, San Nicolas, and Catalina Island. All these specimens were harpooned by commercial swordfish fishermen. At the time, the general opinion of sharks was negative and the killing of these wonderful creatures was considered to be reasonable and even necessary. This focus on white sharks gave Sea World curator Ray Keyes a platform to talk about sharks as important parts of the marine ecosystem which should not be exterminated. In a six year period, Keyes made over 100 television appearances, talked at museums and other venues, hosted a cable television series, guided news and film crews in the field, and contributed to books and magazines (John C. Hewitt, pers. comm. 2009; Ray Keyes, pers. comm. 2009).

A large life-size great white shark model of exceptional quality was recently prepared at the Florida Museum of Natural History, University of Florida, in Gainesville,

Florida. This model was created for a new national traveling exhibition, "Megalodon: Largest Shark That Ever Lived," which was produced by the Florida Museum of Natural History with support from the National Science Foundation. Sculptor Jeff Huber and his collaborators at the Florida Museum of Natural History did their best to make the great white shark model look as real as possible. The shark model was produced by using a whole shark specimen as a plug to create a fiberglass mold. The fiberglass mold was then used to created a fiberglass shark model. The fiberglass shark was tooled and dressed, using bondo filler to sculpt details. Photo references obtained from the University of Florida shark experts were used as a reference when sculpting details. Independently, the teeth were cast from real teeth in urethane and placed in the mouth with epoxy putty. After the sculpture was complete, it was primed, painted and airbrushed to achieve the final finish (Jeff Huber, pers. comm. 2008). The result is without doubt, one of the most beautiful and scientifically precise great white shark models that has ever been produced.

Sculpting Great White Sharks

A less desirable option, which may also work well for educational purposes, is to sculpt a great white shark model from a material, such as foam overlayed with clay. In this case, the entire model is an artist rendering and can be built without a specimen for a model.

Bill Wieger of Kentucky is a sculptor and illustrator of wildlife, fantasy and science-fiction art. He has been able to sculpt some of the best small models of the great white shark that have been produced to date. Bill Wieger has explained to the authors how he has been able to produce impressive work such as one sculpture that portrays the predator as it quietly and elegantly swims.

The first step in creating this sculpture was to compile as much photo reference material as possible. It was extremely helpful that the great whites were viewed and photographed in the wild before starting this project. Photos of every possible angle were obtained so that a complete and scientifically accurate depiction of the shark could be created. A design sketch of the shark was then produced and one-sixth scale (122 cm long) was determined for the size of the sculpture.

An armature and base was then built to support the workpiece. The large wood base (61 × 91 cm) was constructed out of two-by-four lumber and plywood. The armature consisted of a 91-cm tall steel rod that was then attached to the base to which the workpiece would be mounted. Several pieces of foam sheeting were cut and glued together to form the basic shape of the shark and then attached to the steel rod. This foam figure was intentionally constructed oversized and an electric knife was used to carve it into a basic shark shape. Enough foam (about 1") was removed so that when the clay was added, it could be built out to the proper dimensions of the shark sculp-

2—Preserving and Reconstructing Great White Sharks

Life-size sculpture of a 457 cm great white shark attacking a shark cage in the exhibition "Predators," at the Museum of Discovery in Little Rock, Arkansas (photograph courtesy Marci Bynum Robertson, Museum of Discovery, Little Rock, Arkansas).

ture. Clay was then applied over the entire foam core to create the final shape. Various tools and brushes were used to sculpt and refine the clay until it reached the desired shape and texture. It was extremely important to use the photo references as a guide to accurately form all aspects of the shark sculpture. The sculpture is dimensionally accurate. The measurements were obtained from photos and applied to the sculpture. Viewing the shark model from all angles frequently and comparing them to the reference photos was essential to ensure complete accuracy. After the sculpture was completed, the lower jaw and pectoral fins were cut off to be molded separately.

The next step was to create a mold of the clay shark sculpture, which is called creating a "negative" of the clay sculpture. This mold is then used to create a "cast" of the original sculpture. The mold can be made out of several different types of material. In this case, the mold was made out of silicon rubber and fiberglass. Silicone rubber is a thick, fluid material that after curing becomes a flexible solid material; it was used to create the "negative" part of the mold. Because silicone rubber is flexible, a rigid outer jacket of fiberglass was used to help keep the silicone rubber mold in its original shape. Silicone rubber was applied to the clay sculpture in layers, with the first few layers kept thin to ensure that every surface detail would be picked up in the mold. This was an extremely important step and had to be done with care. After two to three layers were applied, subsequent layers were then thickened up with an agent called cabosil and

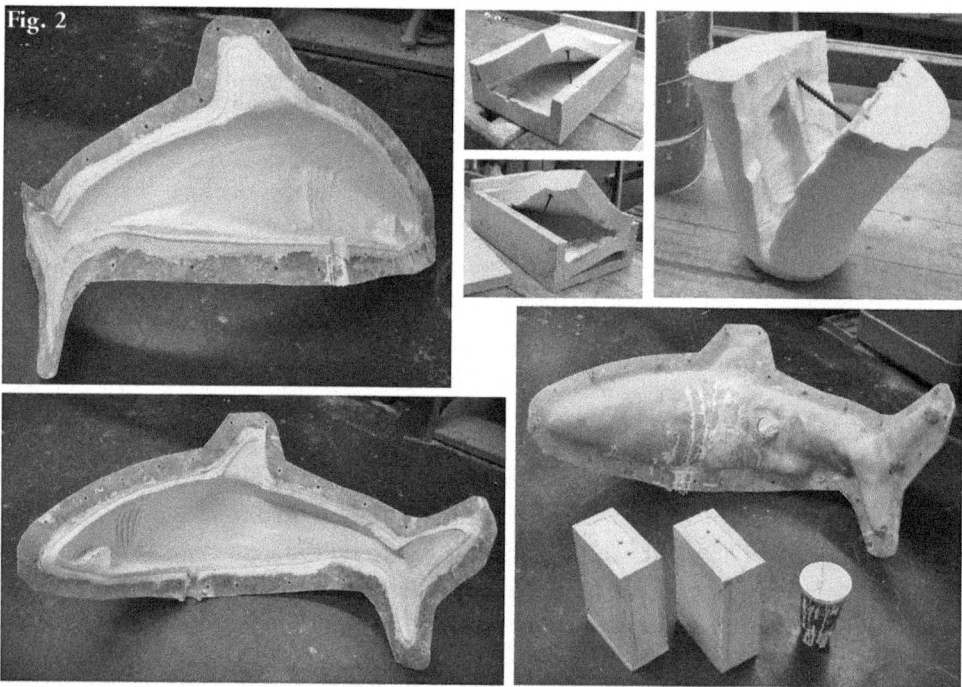

Two steps of the preparation of a 122 cm long sculpture of a great white shark made by artist Bill Wieger (photograph courtesy Bill Wieger). Figure 1 — The finished first model made of foam covered with clay. Figure 2 — The molds for the sculpture (photograph courtesy Bill Wieger). *Following page:* Figure 3 — The completed sculpture made of resin, prepared via cast from the first model of foam and clay (courtesy Bill Wieger).

Fig. 3

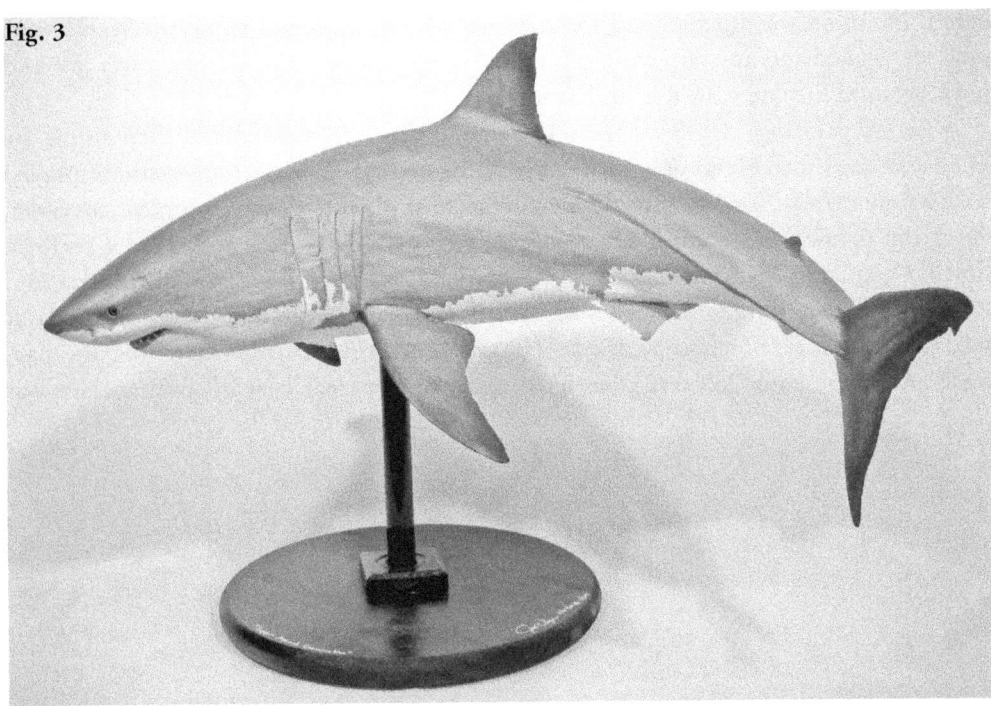

applied in a thickness of one to two inches. After the silicone cured, a center line was drawn defining the mold parting plane on the silicone to differentiate the two sides. A clay dividing wall was then built up along the marked line so that it created a flange perpendicular to the sculpture. Three to four layers of fiberglass were applied over the silicone-covered sculpture and clay dividing wall on only one side of the shark mold. After the fiberglass cured, the clay wall was cleaned off and the opposite side was completed in the same fashion. After the fiberglass cured, several holes were drilled into the fiberglass flange part of the fiberglass jackets, so that bolts could be added later to hold the two sides together. Both fiberglass jackets were then removed from the silicone-covered shark sculpture. The silicone was cut down the marked center line creating two separate halves. Both halves of the silicone mold were removed from the clay sculpture and fitted back into their corresponding fiberglass jackets. Both halves were then cleaned, and bolted together, forming the final mold for casting. The lower jaw and pectoral fins were molded and cast separately using the same process.

Casting is the process of creating a duplicate of the original sculpture. A two-part epoxy resin was used to create the casting. A small hole was drilled in the mold to create a portal where the resin was poured. The two part liquid resin was mixed together in the appropriate quantity and then poured into the mold, partially filling it. The portal was sealed and then the entire mold was strapped onto a rotocasting machine which turned and tumbled the mold until the resin cured. Several layers of resin were built up inside the mold to produce a durable casting. After the last layer of resin completely cured, the mold was opened and the raw cast of the shark was removed. Seams were sanded, cleaned and any imperfections on the casting were repaired. The lower jaw and both pectoral fins are cast in a similar manner.

Before painting, the lower jaw and pectoral fins are glued on and their seams are filled and sanded to blend in with the rest of the casting. The casting was then sprayed with a gray primer. Two to three light coats were applied to ensure complete coverage. After the primer was completely dried, acrylic paint was used to create a realistic finish. Opaque colors were applied to the shark to create the basic overall colorization. Refining and detailing with washes of color fashioned realism and depth to the paint job. Again, photo references were essential as a blue print for color matching and patterns. After the paint job was completed and dried, a clear coat of spray sealant was

Some steps of the preparation of a 1,036 cm long megatooth shark *Carcharodon megalodon* model for the exhibition "Fossil Mysteries" at the San Diego Natural History Museum, San Diego, California. Figure 1 — Small models of the huge megalodon sculpted by Jim Melli, the exhibit preparator and artist (photograph courtesy Jim Melli, San Diego Natural History Museum). *Following pages:* Figure 2 — Full size foam shark constructed by Atomic Props of St. Paul, Minnesota (photograph courtesy S. Campbell, Science Museum of Minnesota). Figure 3 — The teeth that were cast from megalodon shark fossil teeth, being positioned in the model (photograph courtesy S. Campbell, Science Museum of Minnesota). Figure 4 — The shark body that was cut into pieces arrives at the San Diego Natural History Museum (photograph courtesy Lollo Enstad, San Diego Natural History Museum). Figure 5 — The shark being lifted into position (photograph courtesy Tim Murray, San Diego Natural History Museum). Figure 6 — The shark "swimming" in the atrium of the San Diego Natural History Museum (photograph courtesy Tim Murray, San Diego Natural History Museum).

2 – Preserving and Reconstructing Great White Sharks

Fig. 2

Fig. 3

Fig. 4

Fig. 5

Fig. 6

applied. A display base was created out of wood and a PVC pipe was used for mounting, and finished to suit. The completed shark was then attached to the base and ready for presentation.

The museums, institutes and private collectors that desire to commission a great white shark or other marine life sculptures from Bill Wieger, can contact the artist at the following address: Bill Wieger—127 Branch Ct., Shepherdsville, KY 40165; 502-995-3099; Email: billwieger@att.net; Website: billwieger.com

"Fossil Mysteries" is a highly interactive exhibition that opened July 1, 2006, at the San Diego Natural History Museum. The exhibit explores evolution, extinction, ecology, and earth processes. Abundant fossils, models, murals, and dioramas offer multi-sensory experiences. A full-sized model of a 1036 cm long megatooth shark *Carcharodon megalodon* was prepared for this exhibition, which is probably the most accurate depiction ever created of this shark. Michael D. Gottfried, Curator of Vertebrate Paleontology at the Michigan State University Museum, helped Jim Melli, the exhibit preparator and artist, and his team reconstruct the shark. In order to reconstruct the teeth, Melli and his team used casts of an almost-complete set of fossil teeth found in Florida that belonged to one individual megalodon. Based on the size of the teeth they used, Gottfried calculated that the megalodon shark was about 1036 cm long. Based on white shark morphology, Gottfried believes that the megalodon was not as slender as the great white shark, because in general, large animals show a larger girth than smaller animals in order to support all their weight. Also, to be stable, they had wider pectoral fins relative to body size and a shorter snout than a white shark. Jim Melli sculpted a small model of the huge megalodon. One of the museum preparators made a rubber mold from the model of the megalodon and a resin cast was made. Atomic Props of St. Paul, Minnesota, one of the subcontractors working for the Science Museum of Minnesota, constructed the full size shark. The small model, about 30 cm long, made by the San Diego Natural History Museum was laser scanned to create a three dimensional computer file that was scaled up to create the full size shark. Using the computer model, the shark was sliced into 30 cm sections to create foam ring patterns. The foam rings were then cut and assembled to make the full size foam shark. A structural steel frame was designed and built to support the shark in position at the museum. The foam segments were assembled onto the frame, carved and sanded to the final shape under the direction of the museum staff. Once the shape was finalized, the foam core was covered with fiberglass and epoxy resin to form a hard shell. The teeth were cast from fossils found from a megalodon shark, and positioned by shark experts. The final surface details, skin texture and paint colors were completed with the help of museum staff and shark experts. Jim Melli traveled to Minnesota three times to consult on the painting, texture, and color, but the rest of the time he depended on phone conferences and e-mailed pictures. For the final finish, a waterbased epoxy coating was used as a texture coat to get the fiberglass model to look like a real animal. A layer was applied over the fiberglass and smoothed out to eliminate the unnatural, sanded fiberglass look, and also for sculpting details around the eyes, nasal flaps and mouth. After completion, the model of the megalodon body was disassembled in order to ship it across the country to the museum. The shark body was cut into three main pieces, and the fins were also removed. The pieces were then wrapped and shipped from St. Paul, Minnesota, to San Diego, California. At the museum, staff from Atomic Props re-assembled the shark, patched the seams and applied a final coat of paint. Then the shark was lifted into position and bolted in place. Some final detail work was done in position to complete the project. The shark is still swimming in the atrium of the San Diego Natural History Museum (De Rosier, 2006; San Diego Natural History Museum, 2006; Dave Varley, pers. comm. 2008).

3

Great White Shark Materials in Museums

ALASKA

Alaska State Museum
395 Whittier Street, Juneau, AK 99801-1718
Specimens: 1.

The Alaska State Museum in Juneau, Alaska, has in its catalogue materials belonging to one specimen of great white shark. These materials are labeled with the following catalogue number: II-B-981.

Catalogue no. II-B-981 is a pair of teeth, which were preserved dry. Sharks were taken by the Tlingit, who are the northernmost of the Northwest Coast peoples, ranging from southern Alaska to the coast of Oregon. They lived traditionally by fishing and hunting marine animals and built large plank houses, totem poles, and ocean-going dugout canoes. To this day, the livelihood of the Tlingit continues to be linked to the bounty of the natural world. The people maintain interest in both fishing and forestry. Earrings made of large shark teeth were worn by high-ranking Tlingit men during ceremonies. "Shax'-dax' ooxu" is the Tlingit term for shark tooth earrings (De Maddalena et al., 2007).

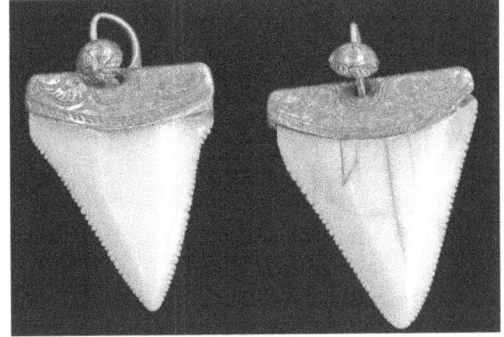

A pair of earrings made from great white shark teeth, preserved at the Alaska State Museum in Juneau, Alaska, with cat. no. II-B-981. These earrings were made from large shark teeth and were worn by high-ranking Tlingit men during ceremonies (photograph courtesy Alaska State Museum).

Sheldon Jackson Museum
104 College Drive, Sitka, AK 99835
Specimens: 1.

The Sheldon Jackson Museum in Sitka, Alaska, has in its catalogue materials belonging to one specimen of great white shark. These materials are labeled with the following catalogue numbers: SJ-I-A-84.

Catalogue no. SJ-I-A-84 is a pair of teeth, which was preserved dry (Rosemary J. Carlton, pers. comm. 2007).

CALIFORNIA

Museum of Vertebrate Zoology, University of California

3101 Valley Life Sciences Building, University of California, Berkeley, CA 94720-3160
Specimens: 1.

The Museum of Vertebrate Zoology of the University of California, in Berkeley, California, has in its catalogue materials belonging to one specimen of great white shark. These materials are labeled with the following catalogue numbers: MVZ 13062 (Milton Hildebrand Catalog 1762).

Catalogue no. MVZ 13062 (Milton Hildebrand Catalog 1762) is a photo of a great white shark that was captured in 1973, near Loreto, Mexico, Baja California Sur (26° 0' 6" N 111° 20' 58" W). This specimen measured 487.7–548.6 cm (16–18 ft) TL. This item is listed on the card in the catalog as "Photo of Jaws w/ Human for Scale." It is a picture of a man on a beach holding the cleaned jaws in front of him. The Museum of Vertebrate Zoology does not have the actual jaws. On the photograph there is the name of Sanford Tanaka, of the Department of Zoology, University of California, Davis. It is unclear whether the photograph was taken by this person, or is a photograph of the person. The determiner was Milton Hildebrand (Chris Conroy, pers. comm. 2008, 2009; Craig Moritz, pers. comm. 2008; Ichthyology Collection Database of the Museum of Vertebrate Zoology in Berkeley, 2008).

Natural History Museum of Los Angeles County

900 Exposition Boulevard, Los Angeles, CA 90007
Specimens: 27 (including 2 missing).

The Natural History Museum of Los Angeles County in Los Angeles, California, has in its catalogue materials belonging to 27 specimens of great white shark. These materials are labeled with the following catalogue numbers: one item without catalogue number, 35875-1, 35892-1, 35892-2, 35893-1, 37513-1, 38194-1, 39276-19, 39341-1, 39431-1, 39474-1, 39475-1, 39475-2, 42094-1, 42094-2, 42100-2, 42133-1, 42726-1, 42728-1, 42894-1, 43638-1, 43800-1, 43804-1, 43805-1, 44842-1, 56304-15, 56731-1.

A model of a 190 cm white shark and one small set of jaws were on exhibit in the Marine Hall, but they were recently removed while the hall is being renovated. Perhaps the model will be reinstalled once the new hall is complete (Jeffrey A. Seigel, pers. comm. 2009).

Item without catalogue number is a whole specimen. The great white shark was captured in 1976, off Ventura County, California. It was free-swimming and had become entangled in Ben Henke's gill net. This specimen measured 81.3 cm (32 in) TL. It was the smallest confirmed free swimming white shark known to science. The collector was commercial fisherman Ben Henke, via Ralph S. Collier. This specimen has not been found, and the reason it is missing is unknown. Some time ago, a search was done by Ralph S. Collier and Camm Swift, but they could not find the specimen in the Natural History Museum of Los Angeles County collection. The item is not currently on exhibit to the public (Ralph S. Collier, pers. comm. 2008).

The items with catalogue no. 35875-1 are fins, some teeth and vertebral column, which were preserved

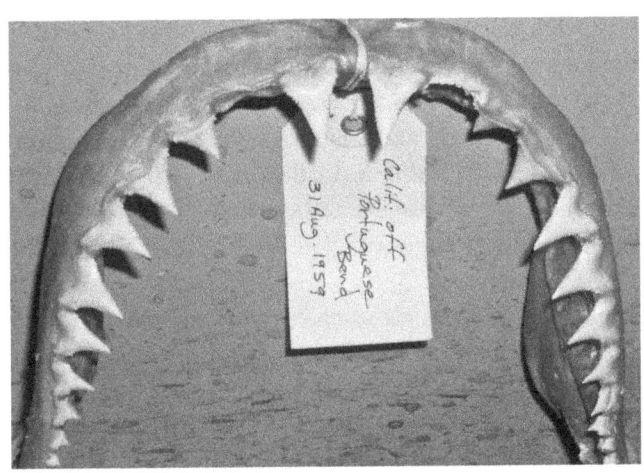

Close-up of the upper jaw of a set of jaws from a great white shark that was captured on August 31, 1959, off Portuguese Bend, California, and preserved at the Natural History Museum of Los Angeles County in Los Angeles, California, with cat. no. 39474-1 (photograph courtesy Antonella Preti, Natural History Museum of Los Angeles County).

dry. The teeth were inadvertently cut out of jaws, but the complete tooth set from right side was reconstructed by Ralph S. Collier and Shelton P. Applegate. The fins, teeth and vertebral column were from a great white shark that was captured on June 28, 1976, off Ventura, about 17 miles NNW Santa Barbara Is., California. This specimen measured 507.9 cm TL and weighed 2152.3 kg (4745 lb), to date a record weight for a white shark in the state of California. The weight was obtained from the California Department of Weights and Measures truck scale house in Camarillo, California. The weight may exceed that of other record sharks. As for the scale, it was checked and found to be accurate to within plus or minus 16 ounces. There were no contents in the stomach. The collector was Hans Weeren aboard the vessel *Commanche*. Considerable data and photographs for this specimen are on file in the Natural History Museum of Los Angeles County. The item is not currently on exhibit to the public (Ralph S. Collier, pers. comm. 2008; Jeffrey A. Seigel, pers. comm. 2008).

Catalogue no. 35892-1 are vertebrae, which were preserved in liquid. The vertebrae were from a great white shark that was captured on July 14, 1976, off Ventura, off the mouth of the Ventura River and Pierpoint Bay, California. The depth of water was 15 m, and the depth of capture was between 4 and 11 m. This specimen measured 225.8 cm TL. The shark was captured by commercial fisherman Ben Henke, and then given to Camm Swift, Ralph S. Collier and Katherine Binyon, who then transferred it to the Natural History Museum of Los Angeles County. The item is not currently on exhibit to the public (Jeffrey A. Seigel, pers. comm. 2008, 2009).

Catalogue no. 35892-2 are vertebrae, which were preserved in liquid. The vertebrae were from a great white shark that was captured on July 14, 1976, off Ventura, off the mouth of the Ventura River and Pierpoint Bay, California. The depth of water was 15 m, and the depth of capture was between 4 and 11 m. This specimen measured 150.0 cm TL. The shark was captured by commercial fisherman Ben Henke, and then given to Camm Swift, Ralph S. Collier and Katherine Binyon who then transferred it to the Natural History Museum of Los Angeles County. The item is not currently on exhibit to the public (Jeffrey A. Seigel, pers. comm. 2008, 2009).

Catalogue no. 35893-1 is a single vertebra, which was preserved in liquid. The vertebra was from a great white shark that was captured on July 27, 1976, off Pietas Pt., approximately 4 mi N Ventura, California. The depth of water was 26 m, and the depth of capture was 7 m. This specimen measured 136.5 cm TLn and 141.0 cm TOT. The collector was commercial fisherman Ben Henke. The item is not currently on exhibit to the public (Jeffrey A. Seigel, pers. comm. 2008, 2009).

Catalogue no. 37513-1 is a set of jaws, which was preserved dry. The jaws were from a great white shark that was captured in July 1963, in Santa Monica Bay, Paradise Cove, California. The collector was Shelton P. Applegate. The item is not currently on exhibit to the public (Jeffrey A. Seigel, pers. comm. 2008).

Catalogue no. 38194-1 is a set of jaws, which was preserved dry. The jaws were from a great white shark that was captured approximately 60 miles off San Diego, California, at 32° 5' N 118° 14.5' W, probably in a spot called the "sixty mile bank." The depth of capture was between 0 and 97 m. This specimen measured 340 cm TL. This item was acquired by the Natural History Museum of Los Angeles County on December 14, 1978. It was purchased in Scottsdale, Arizona, and received by Ralph S. Collier. The owner and donor was Steve Pruitt. According to Ralph S. Collier, some of the data for this specimen is questionable; the length of the shark might have been an estimate. The item is not currently on exhibit to the public (Jeffrey A. Seigel, pers. comm. 2008).

Catalogue no. 39276-19 is a tooth set, which was preserved dry. The item is not currently on exhibit to the public (Jeffrey A. Seigel, pers. comm. 2008).

Catalogue no. 39341-1 is a set of jaws, which was preserved dry. This item was acquired by the Natural History Museum of Los Angeles County in April-May 1965. The item is not currently on exhibit to the public (Jeffrey A. Seigel, pers. comm. 2008).

Catalogue no. 39431-1 is a set of jaws, which was preserved dry. The jaws were from a great white shark that was captured in 1955-1956, off La Jolla, near Scripps Pier, California. This item is not currently on exhibit to the public (Jeffrey A. Seigel, pers. comm. 2008).

Catalogue no. 39474-1 is a set of jaws, which was preserved dry. The jaws were from a great white shark that was captured on August 31, 1959, off Portuguese Bend, California. This item is not currently on exhibit to the public (Jeffrey A. Seigel, pers. comm. 2008).

Catalogue no. 39475-1 is a skin, which was preserved dry. The skin was from a great white shark that was captured on August 23, 1962. The item is not currently on exhibit to the public (Jeffrey A. Seigel, pers. comm. 2008).

Catalogue no. 39475-2 is a pair of claspers. The claspers are from a male great white shark that was captured on August 23, 1962. This item is not currently on exhibit to the public (Jeffrey A. Seigel, pers. comm. 2008, 2009).

Catalogue no. 42094-1 is a vertebra, which was preserved dry. The vertebra was from a great white shark that was captured in April 1965, off Baja, from a Beach 4 mi. north of Oakie's Landing (S of Puertecitos), Mexico. The item is not currently on exhibit to the public (Jeffrey A. Seigel, pers. comm. 2008).

Catalogue no. 42094-2 is a vertebral column, which was preserved dry. The vertebral column was from another great white shark that was also captured in April 1965, off Baja from a beach 4 mi. N of Oakie's Landing (S of Puertecitos), Mexico. The item is not currently on exhibit to the public (Jeffrey A. Seigel, pers. comm. 2008).

Catalogue no. 42100-2 is a vertebral column, which was preserved dry. The vertebra was from a great white shark that was captured off Sandy Hook, New Jersey. The item is not currently on exhibit to the public (Jeffrey A. Seigel, pers. comm. 2008).

The items with catalogue no. 42133-1 are two vertebrae, which were preserved dry. The vertebrae were from a great white shark that was captured on August 5, 1964, ¼ mile east of Highlands Bridge, Florida (this locality is doubtful). This specimen measured 143.5 cm TL. The item is not currently on exhibit to the public (Jeffrey A. Seigel, pers. comm. 2008).

Catalogue no. 42726-1 is a skeleton, which was preserved dry. The skeleton is from a great white shark that was captured in September 1977, off Ventura, California. This specimen measured 209.9 cm TL. The collector was R.W. Huddleston. The item is not currently on exhibit to the public (Jeffrey A. Seigel, pers. comm. 2008).

Catalogue no. 42728-1 is a set of jaws, which was preserved dry. The jaws were from a great white shark that was captured on May 20, 1982, 16–20 miles off Point Dume near Santa Barbara Island (Loran reading – 41250 & 28050), California. The depth of capture was near the surface between 0 and 6.1 m. The conditions at capture were: air temperature was 61°F, water temperature was 61–62°F, and weather was cloudy. This specimen measured 460.9 cm TL and weighed slightly over 1224.7 kg (2700 lb). The collector was Edward Peters. This item is not currently on exhibit to the public (Jeffrey A. Seigel, pers. comm. 2008, 2009).

The items with catalogue no. 42894-1 are dorsal fins, pectoral fins and vertebrae. The fins and vertebrae were from a great white shark that was captured on August 30, 1982, 5 miles off Point Dume, California. The depth of capture was between 0 and 9.15 m. It was caught in a drift gill net set at 3 fathoms. The net was set at 10 P.M. on August 29 and retrieved at 6 A.M. on August 30, 1982. The shark was still alive when caught. This specimen measured 494.2 cm TL and its weight was estimated at 680.4–907.2 kg (1500–2000 lb). The liver weighed 272.2–317.5 kg (600–700 lb) (422.3 kg – 931 lb – gross weight, with metal box estimated at 90.7 kg, weighed on scales at

Pioneer Fish Co., San Pedro). This shark had one elephant seal head in its stomach (it was likely a northern elephant seal *Mirounga angustirostris*). The collector was skipper Craig Williams aboard the fishing vessel *Procalo*. The item is not currently on exhibit to the public (Jeffrey A. Seigel, pers. comm. 2008, 2009).

Catalogue no. 43638-1 was from a great white shark that was captured on May 19, 1984, off barge Isle of Redondo, 1.5 mi. offshore west of Redondo, California. The depth of water was 33.5 m, and the depth of capture was 0 m. This specimen measured 132.1 cm TL and weighed 18.1 kg (40 lb). The collectors were Gordon King, Robert Moughalian and Mark Willey. It was cataloged as being some anatomical parts. These parts have not be found and the reason they are missing is unknown. Data and photos only have been found. The item is not currently on exhibit to the public (Jeffrey A. Seigel, pers. comm. 2008, 2009).

The items with catalogue no. 43800-1 are vertebrae, claspers and pelvic fins, heart, testes, stomach contents (possibly California sea lion *Zalophus californianus* remains). The vertebrae, claspers, pelvic fins, heart and testes were from a male great white shark that was captured on May 15, 1985, 8.5 miles northwest of the west end of Catalina Island, California (33° 36' N 118° 37' W), in 580 m deep water. The surface water temperature at the time of capture was 59°F. The drift gill net was 1000 fms long of mostly 16–18 inch mesh. This specimen measured 435.3 cm TL. The collector was Ramiro Soares aboard the vessel *Ana Maria*. This item is not currently on exhibit to the public (Jeffrey A. Seigel, pers. comm. 2008, 2009).

Catalogue no. 43804-1 is a whole specimen, which was preserved in liquid. The male great white shark was captured on June 12–13, 1985, approximately 0.5 miles southwest of Newport Pier, California. The depth of water was 13 m. The water temperature was 67°F. The weather was warm and calm and there were many fish in the water, including anchovies and barracuda. Six different sharks and rays were caught in the net, including two great white sharks (for the second specimen see catalogue no. 43805-1) and tope sharks *Galeorhinus galeus* (Linnaeus, 1758). It was likely a set gill net: there was a set net fishery for tope shark that was banned in the 1990s. This specimen measured 126.1 cm TL. The collector was Randy Shelton. The item is not currently on exhibit to the public (Jeffrey A. Seigel, pers. comm. 2008, 2009).

The items with catalogue no. 43805-1 are a set of jaws, vertebrae (2 were removed), and stomach contents. The jaws and vertebrae were from a great white shark that was captured on June 12–13, 1985, approximately 0.5 miles southwest of Newport Pier, California. The depth of water was 13 m. It was caught in the net together with the specimen cited above (catalogue no. 43804-1, see above for additional data). This specimen measured 130.7 cm TL. The item is not currently on exhibit to the public (Jeffrey A. Seigel, pers. comm. 2008, 2009).

The items with catalogue no. 44842-1 are a set of jaws, vertebrae, skin and tissue. The jaws, vertebrae, skin and tissue were from a great white shark that was captured on January 3, 1990, 300 yds off Sunset Beach in Santa Monica Bay, California. The

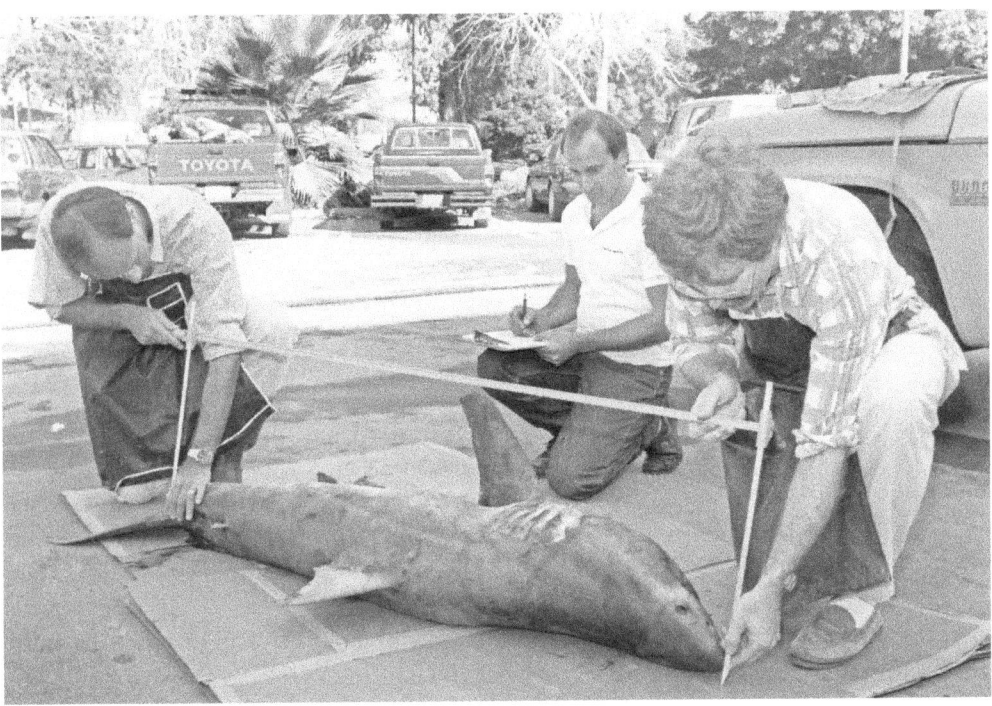

A 184.3 cm great white shark captured on January 3, 1990, off Sunset Beach in Santa Monica Bay, California. The jaws, vertebrae, skin and tissue from this specimen are preserved at the Natural History Museum of Los Angeles County in Los Angeles, California, with cat. no. 44842-1 (photograph courtesy Natural History Museum of Los Angeles County).

depth of water was 7.6 m. The water temperature was 62°F and the water was murky with kelp. The weather was clear. This specimen measured 184.3 cm TL. The collector was Mr. Stephen Daratsos, Jr. The skin was used in a shark exhibit, but the item is not currently on exhibit to the public. Here is a full report of the capture in the words of Stephen Daratsos, Jr., the angler who caught this white shark.

> The night before the catch, the local TV news reported a white shark catch by some famous New York baseball player. I called some friends asking if they were interested in trying to catch one. They all declined so I went fishing by myself. I was not sure of the water depth so I started in deep water as I am accustomed to catching the pelagic varieties such as makos. I spent most of the day in the deep water. As the sun was going down, I thought I would try very shallow water. I used mackerel for bait but no chum or any other attractant. I was quite close to the shore and did not expect to have action there since it is a popular Los Angeles area surfing beach. I just did not expect to find white sharks that close to surfers. I let out my line, put the rod in a holder, and sat down at the helm of my drifting boat while fiddling with my wristwatch. About one and a half minutes went by before I heard the reel's clicker start to make noise. Then, the reel started peeling out line at a rapid rate. I set the hook and the fight was on. Suddenly, the line went limp as the shark was able to free the hook from its mouth. I left the line out and within seconds, the shark took the bait again. This time I set the hook harder. The reel I was using was a Daiwa 450H on a Fenwick Pacificstick with a 30 to 80 lb rating. I had on

80 lb monofilament line and a custom wire leader. As the reel started to get near the end of the line, I knew if I did nothing, I would lose the fish. I started the engine and put the boat into reverse gear. I held the rod in one hand and the steering wheel in the other. About 45 minutes later, I had the fish worn down. When I got the fish to leader, I had to gaff it myself as I was the only person on board. First, I tied a noose in a dock line and attached it to a boat cleat. Then, I waited for the shark to pass under the boat again and I gaffed the tail area. I threw the fishing rod down on the deck and held onto the gaff as the shark thrashed around. I figured that as long as I had the tail out of the water, I could eventually get the upper hand. When the shark quieted down a bit, I slipped the noose over the tail, took up the slack & cleated the rope. I waited for the shark to stop moving before doing anything else. I drove slowly back to the dock where it was now dark. There was hardly anybody there and then people started appearing. First it was passers-by, then the county sheriffs, then the local ABC News showed up. It was seen on the local news that night and replayed again in the morning. As I drove home trailering the boat, drivers on the freeway were all trying to get behind me to see that shark as I left it tied onto the swim step. In the morning, I thought about what to do with it. The Los Angeles Museum of Natural History came to my mind as a possible place to donate it. I transferred the shark to the back of my pickup truck and covered it with a polyethylene tarpaulin. When I arrived at the museum, Jeffrey Siegel and some other experts went outside to measure and take samples from the shark. I felt that I did the right thing to donate my catch in the name of research and education.

Stephen Daratsos, Jr., is now on the side of conservation. Beginning in 1994, the great white shark became a protected species in California waters (see page 81, "Conservation") (Stephen Daratsos, Jr., pers. comm. 2008; Jeffrey A. Seigel, pers. comm. 2008, 2009).

Catalogue no. 56304-15 is a set of jaws, which was preserved dry. The item is not currently on exhibit to the public (Jeffrey A. Seigel, pers. comm. 2008, 2009).

The items with catalogue no. 56731-1 are vertebrae, which were preserved in liquid (70 percent ethyl alcohol). The jaws were from a female great white shark that was captured on November 14, 2000, off the east end of Catalina Island, California. The fishing vessel was fishing for sardines. It was likely a purse seiner. The great white shark was likely taken incidentally during a wrap. The shark was landed at 22nd Street, J&D Seafood, at 11:00 A.M. This specimen measured 433.0–449.1 cm TL. The liver weighed 129.3 kg (285 lb). The collector was Frank Colona, aboard the vessel *Santa Maria*. The item is not currently on exhibit to the public (Jeffrey A. Seigel, pers. comm. 2008).

Moss Landing Marine Laboratories

8272 Moss Landing Road, Moss Landing, California 95039-9619
Specimens: 12 (including 2 missing).

The Moss Landing Marine Laboratories in Moss Landing, California, has in its catalogue materials belonging to 12 specimens of great white shark. These materials are labeled with the following catalogue numbers: without catalogue number, B-33-0305, B-34-0305, B-35-0305.

3 — Materials in Museums — California

Roger C. Helm performing a necropsy of a 472.4 cm TL female great white shark that was found beached near Año Nuevo State Park, off the coast of central California, in February 1977, of which a dorsal fin is preserved at the Moss Landing Marine Laboratories, Moss Landing, California with cat. no. B-33-0305 (photograph courtesy Roger C. Helm, Moss Landing Marine Laboratories).

Catalogue no. B-33-0305 is a dorsal fin taken off a female great white shark that was found beached near Año Nuevo State Park, off the coast of central California, in February 1977. This specimen measured 472.4 cm (15.5 ft) TL and 365.8 cm (12 ft) in circumference at her pectoral fins. The collector and the determiner was Roger C. Helm. The item is kept in a bin for teaching. Its stomach contained about 10 large barely digested pieces from a subadult male northern elephant seal *Mirounga angustirostris*. The pieces that Roger C. Helm remembers include the head back to the third cervical vertebra, the hind flippers completely intact, one fore flipper including the shoulder blade, and several approximately 40 cm diameter chunks of flesh. In total, Helm would estimate the shark had consumed a couple hundred pounds of elephant seal, but he did not measure or weigh any of the pieces. There were no obvious external injuries. The shark was aged by Jim Harvey at Moss Landing Marine Laboratories (Gregor M. Cailliet, pers. comm. 2008; Roger C. Helm, pers. comm. 2008).

Catalogue no. B-34-0305 is a whole specimen. The great white shark was captured on September 24, 1978, 1 mi. offshore from the cement boat off Seacliff Beach, Monterey Bay, California. It was tangled in a trammel net. The collectors were Fred Salter and Jim Miller. This item is currently missing and perhaps it was never kept at the Moss Landing Marine Laboratories in Moss Landing for lack of room (Gregor M. Cailliet, pers. comm. 2008).

Catalogue no. B-35-0305 is a whole specimen. The male great white shark was captured on September 10, 1984, off Bodega Bay, California. It was caught in a gill net. This specimen measured 147.3 cm (4-ft 10-inch) TL and weighed 24.9 kg (55 lb). The collector was fisherman Joe Papetti. This white shark was kept in captivity for 11 days at the Monterey Bay Aquarium, Monterey, California, and died on the day it was planned to be released, September 20 (see page 177, "September 1984: Monterey Bay Aquarium, Monterey, California"). When it died, it was stored in the Monterey Bay Aquarium freezer for some time and then turned over to Gregor M. Cailliet at Moss Landing Marine Laboratories, where it was preserved and stored (it was catalogued perhaps on January 15, 1986). This item is currently missing. From the Bin Fishes inventory list from Ichthyology of 2007, the item appeared to no longer be at Moss Landing Marine Laboratories. The Moss Landing Marine Laboratories suffered serious damage from the 1989 Loma Prieta earthquake. The laboratories were torn down and reconstructed. All the ichthyology specimens were moved to other facilities and the California Academy of Sciences in San Francisco, California, was the only facility large enough to accommodate such a specimen. Therefore, it is thought that this specimen or parts of it are among those currently preserved at the California Academy of Sciences (see page 120, "Moss Landing Marine Laboratories") (Gregor M. Cailliet, pers. comm. 2008; Ellis and McCosker, 1991; Monterey Bay Aquarium, 2008b; Powell, 2001; David C. Powell, pers. comm. 2008).

The items without catalogue numbers are vertebrae, which were preserved frozen. The vertebrae were from nine great white sharks and were taken from the area anterior to the first dorsal fin. Four of these nine specimens measured 234 cm TL, 393 cm TL, 460.9 cm TL, and 507.9 cm TL. All nine great white sharks were caught in commercial gill nets. For each individual specimen, measurements were taken (total length, precaudal length, distance between dorsal fin origins, girth and weight) and the reproductive tract was examined. These items were acquired by the Moss Landing Marine Laboratories during 1978 to 1984. These vertebrae have been used primarily for age, growth, and bomb radiocarbon age validation studies (Cailliet, 1990; Cailliet *et al.*, 1983; Welden *et al.*, 1987). These specimens would be difficult to find and are not accessioned into Moss Landing Marine Laboratories collection (Gregor M. Cailliet, pers. comm. 2008).

Pelagic Shark Research Foundation
750 Bay Ave, #216, Capitola, CA 95010
Specimens: 2.

The Pelagic Shark Research Foundation in Santa Cruz, California, has in its catalogue materials belonging to two specimens of great white shark. These materials are without catalogue numbers.

Callaghan Fritz-Cope (left) and Sean R. Van Sommeran (right) of the Pelagic Shark Research Foundation, Santa Cruz, California, holding the tail from a 433 cm TL female great white shark that was captured on November 23, 2000, in Morro Bay, California. This item (without cat. no.) is owned by the Pelagic Shark Research Foundation and stored at the Marine Wildlife Veterinary Care and Research Center of the California Department of Fish and Game, Santa Cruz, California (photograph courtesy Pelagic Shark Research Foundation).

The first items without a catalogue number are a first dorsal fin and a tail including caudal peduncle, caudal fin, second dorsal fin and anal fin, which were preserved frozen. The first dorsal fin and tail were from a female great white shark that was captured on November 23, 2000, approximately 4.8 km from shore, off Morro Bay, California (35°10' N–120°40' W). It was incidentally caught in a halibut gill net set in water 60 m deep. This specimen measured 433 cm TL and weighed 772 kg. The shark was listed on the internet for sale, illegally. The Pelagic Shark Research Foundation intercepted the item and confiscated it using their Scientific Collecting Permit. The shark was frozen, and then stored at the Marine Wildlife Veterinary Care and Research Center of the California Department of Fish and Game, Santa Cruz, California. The shark was later thawed and David A. Ebert of Moss Landing Marine Laboratories conducted a necropsy of it. There were no contents in the stomach. Five species of copepods (*Dinemoura producta*, *D. latifotlia*, *Echthrogaleus coleoptratus*, *Pandarus bicolor*, and *Achtheinus oblongus*) were collected from the external body surface of the shark. The copepods were fixed in 10 percent formalin or 70 percent ethanol and sent to George W. Benz for identification. All of the specimens were then retained in the personal collection of George W. Benz. The electrosensors and canals of sharks are filled with a uniform hydrogel. Brandon R. Brown, Mary E. Hughes and Clementina Russo have taken samples of this gel from the great white shark caught off Morro Bay, and they concluded that the shark hydrogel strongly localizes ionic species. The first dorsal fin and tail are stored at the Marine Wildlife Veterinary Care and Research Center of the California Department of Fish and Game, Santa Cruz, California. The item is not currently on exhibit to the public (Benz et al., 2003; Brown, 2005; Stephens, 2001; Sean R. Van Sommeran, pers. comm. 2008).

The second item without a catalogue number is a set of jaws, which was preserved dry. This item was acquired by the Pelagic Shark Research Foundation in January 2009. The item is not currently on exhibit to the public, but is used for teaching and education (Sean R. Van Sommeran, pers. comm. 2008).

Scripps Institution of Oceanography, University of California San Diego

8602 La Jolla Shores Drive, La Jolla, CA 92037
Specimens: 11.

The Scripps Institution of Oceanography in La Jolla, California, has in its catalogue, materials belonging to 13 specimens of great white shark. These materials are labeled with the following catalogue numbers: two without catalogue numbers, SIO 04-43, SIO 05-66, SIO 06-78, SIO 45-197, SIO 47-67, SIO 48-194, SIO 55-95, SIO 71-196, SIO 77-72, SIO 88-105, SIO 88-105.

A 365.8 cm (12 ft) white shark replica, recreated from a cast of an original specimen, hangs in the lobby of the Birch Aquarium at the Scripps Institution of Oceanography. This replica was cast by Lyons and O'Haver, Inc., San Diego, California. It is unclear if this replica was cast from one of the specimens that are listed in the Ichthyology Collection Database of the Scripps Institution of Oceanography and described below, or if it is a different specimen.

Catalogue no. SIO 04-43 is a small sample of muscle tissue from a great white shark that was captured on September 10, 2003, off Newport Beach, California (33°31.8' N–117°56.2' W). The depth of water was approximately 91.4 m and the depth of capture was approximately 91.4 m. It was caught incidentally in a set gill net used for halibut. This specimen measured 147.0 cm TL. The collector was a local fisherman (via Chris Lowe, California State University, Long Beach). The item is not currently on exhibit to the public (Ichthyology Collection Database of the Scripps Institution of Oceanography, 2008; H.J. Walker, Jr., pers. comm. 2008).

Catalogue no. SIO 05-66 is a small sample of muscle tissue from a great white shark that was captured on August 8, 2005, off Ventura, California (34°17.0' N–119°18.0' W). It was caught in a gill net on bottom by a commercial fisherman and died before it could be released. This specimen measured 130.0 cm TL and the collector was a fisherman. The item is not currently on exhibit to the public. The specimen was delivered to the laboratory of Chris Lowe in Long Beach, California. Daniel P. Cartamil of the Scripps Institution of Oceanography received the carcass from Chris Lowe in order to do some Magnetic Resonance Imaging (MRI) scans for a muscle morphology study. This animal was imaged in the early days of the Digital Fish Library project. The Digital Fish Library (DFL), funded by the National Science Foundation, is a collaborative project at the University of California, San Diego between the Center for Scientific Computation in Imaging (CSCI), the Center for functional Magnetic Resonance Imag-

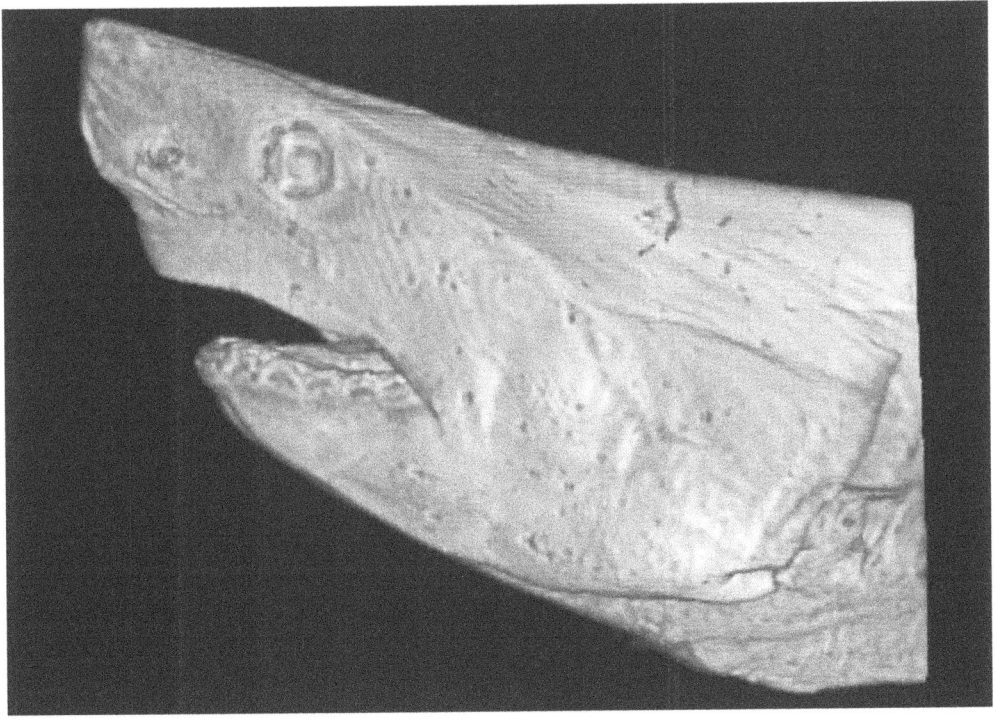

Sagittal magnetic resonance image volume rendering of the head of a 130 cm TL great white shark captured on August 8, 2005, off Ventura, California, and preserved as a whole specimen at the Scripps Institution of Oceanography in La Jolla, California, with cat. no. SIO 05-33. The item is not currently listed in the Scripps Institution of Oceanography collection database, with the exception of a small sample of muscle tissue with cat. no. SIO 05-66 (courtesy Digital Fish Library/Center for Scientific Computation in Imaging/Center for Functional Magnetic Resonance Imaging/Scripps Institution of Oceanography/National Science Foundation).

ing (CfMRI) and the Scripps Institution of Oceanography (SIO), including the Birch Aquarium at Scripps. The DFL mission is to catalog the anatomical MRI data of fishes to provide a resource for research and education. Normally, fish are studied by cutting specimens into slices (cross sections) so that direct measurements, such as muscle locations and volumes, can be made. Such dissections can only be studied by a limited number of researchers, and often destroy the integrity of the fish specimen. However, MRI is non-invasive and allows for more detailed measurements, so it is a scientifically valuable way to collect high resolution 3D images of fishes that can be digitally dissected over and over again by anyone visiting the DFL. Magnetic resonance imaging works by detecting the spatial modulations in the magnetic resonance signals from the protons in hydrogen atoms, which are present in the water of soft tissue, in the presence of spatially varying externally applied magnetic fields. This great white shark specimen was imaged using a three-dimensional spoiled gradient-recalled (3D-SPGR) MRI pulse sequence with a voxel size $1.17 \times 1.17 \times 1.10$ and a slice thickness of 1.1 mm. This scanning was filmed by National Geographic for a *Shark Week* segment (Rachel M. Berquist, pers. comm. 2008; Daniel P. Cartamil, pers. comm. 2008; Digital Fish Library, 2008; Lawrence R. Frank, pers. comm. 2009; Ichthyology Collection Database of the Scripps Institution of Oceanography, 2008; H.J. Walker, Jr., pers. comm. 2008).

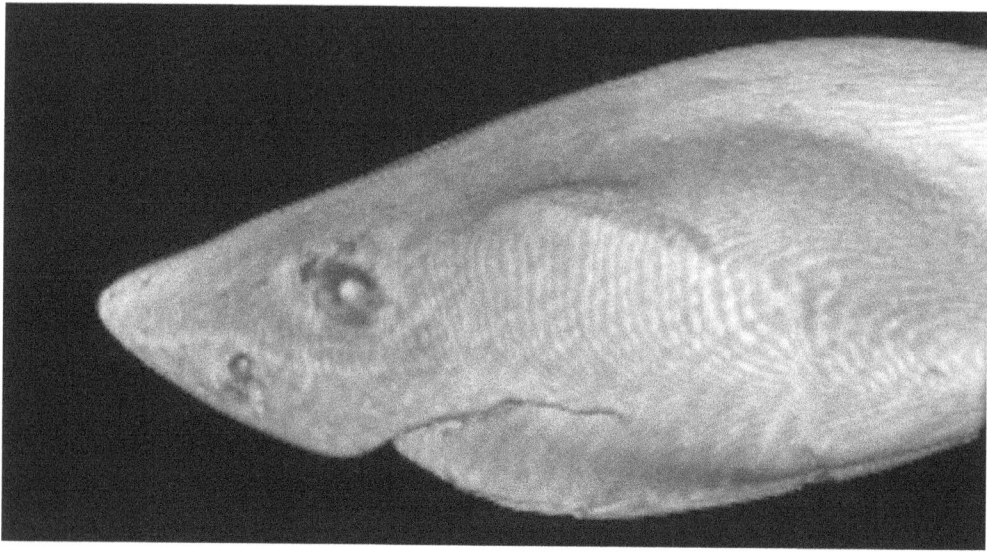

Sagittal magnetic resonance image volume rendering of the head of a great white shark captured on June 13, 2006, off Long Beach, California, and preserved at the Scripps Institution of Oceanography in La Jolla, California, with cat. no. SIO 06-78. The item is not currently listed in the Scripps Institution of Oceanography collection database, with the exception of a small sample of muscle tissue (courtesy Digital Fish Library/Center for Scientific Computation in Imaging/Center for Functional Magnetic Resonance Imaging/Scripps Institution of Oceanography/National Science Foundation).

Catalogue no. SIO 06-78 is a small sample of muscle tissue from a great white shark that was captured on June 13, 2006, off Long Beach, California (33°45.0' N–118°11.0' W). It was caught in a gill net. In mid–June 2006, this great white shark was also MR imaged by the Digital Fish Library. The item is not currently on exhibit to the public (Rachel M. Berquist, pers. comm. 2008; Ichthyology Collection Database of the Scripps Institution of Oceanography, 2008; H.J. Walker, Jr., pers. comm. 2008).

The items without catalogue numbers are dermal denticles, which were preserved dry. These items are not currently on exhibit to the public (H.J. Walker, Jr., pers. comm. 2008).

The item without catalogue number is a vertebra, which was preserved dry. The vertebra was from a great white shark that was captured on June 13 or 14, 1976, off Santa Catalina Island, California. This specimen measured 549.9 cm TL and weighed 1882 kg. The item is not currently on exhibit to the public (H.J. Walker, Jr., pers. comm. 2008).

Catalogue no. SIO 47-67 was from a great white shark that was observed by a local fisherman on April 13, 1947, North of San Felipe, Baja California, Mexico (31°00.0'N, 114°52.0'W) The item is not currently on exhibit to the public. Regarding this item, the database of the Scripps Institution of Oceanography says "specimen non saved," but it is unclear what it means (Ichthyology Collection Database of the Scripps Institution of Oceanography, 2008; H.J. Walker, Jr., pers. comm. 2008).

Catalogue no. SIO 48-194 was from a female great white shark that was captured on July 21, 1948, near buoy boat anchorage off Scripps Institution of Oceanography Pier in San Diego, California. (32°52.0'N, 117°15.0'W). It was caught by hook and line. The collector was B. Ramsower. The item is not currently on exhibit to the public (Ichthyology Collection Database of the Scripps Institution of Oceanography, 2008; H.J. Walker, Jr., pers. comm. 2008).

The items with catalogue number SIO 55-95 are a set of jaws, vertebrae and skin, which were preserved dry. The jaws, vertebrae and skin were from a great white shark that was captured on November 3, 1955, at the end of Scripps Institution of Oceanography Pier in San Diego, California (32°52.0'N, 117°15.0'W). The depth of water was 6.1 m and the depth of capture was 3 m. It was caught on a setline. The database of the Scripps Institution of Oceanography reports two lengths for this specimen, 180.6 cm and 162.6 cm TL. It is unclear what it means. The different lengths may refer to different measurements, such as TOT and TLn, or maybe the items labeled with catalogue number SIO 55-95 come from two different specimens, even though this seems unlikely. The collectors were A.O. Flechsig & party. These items are not currently on exhibit to the public (Ichthyology Collection Database of the Scripps Institution of Oceanography, 2008; H.J. Walker, Jr., pers. comm. 2008).

Catalogue number SIO 71-196 was from a female great white shark that was captured on November 4, 1971, off La Jolla Cove, California. (32°51.0'N, 117°16.0'W). The depth of water was 15.2 m and it was caught in a lobster trap, presumably entangled.

This specimen measured 200.0 cm TL and the collector was D. Tomlinson. The item is not currently on exhibit to the public (Ichthyology Collection Database of the Scripps Institution of Oceanography, 2008; H.J. Walker, Jr., pers. comm. 2008).

Catalogue no. SIO 77-72 was from a great white shark that was captured on September 16–17, 1976, three miles off Dana Point, California (33°27.0'N, 117°37.0'W). The depth of capture was 35 fm and it was caught in a gill net. This specimen measured 158.0 cm TL. The collector was T. Magill aboard the vessel MV *Two Sons*. The item is not currently on exhibit to the public (Ichthyology Collection Database of the Scripps Institution of Oceanography, 2008; H.J. Walker, Jr., pers. comm. 2008).

Catalogue no. SIO 88-105 was from a male great white shark that was captured on July 14, 1988, off Black's Beach near Bathtub Rock just north of Scripps Pier in San Diego, California (32°53.0'N, 117°15.0'W). The depth of water was 12.2 m and it was caught in a gill net. This specimen measured 144.0 cm TL and the collector was T. Lewis. The item is not currently on exhibit to the public (Ichthyology Collection Database of the Scripps Institution of Oceanography, 2008; H.J. Walker, Jr., pers. comm. 2008).

The items with catalogue number SIO 88-105 are a set of jaws, which were preserved dry, and vertebrae, which were preserved in liquid (isopropyl alcohol). The jaws and vertebrae were from a female great white shark that was captured on July 14, 1988, off Black's Beach near Bathtub Rock just north of Scripps Pier in San Diego, California (32°53.0'N, 117°15.0'W). The depth of capture was 12.2 m and it was caught in a gill net. This specimen measured 163.0 cm TL and the collector was T. Lewis. The item is not currently on exhibit to the public (Ichthyology Collection Database of the Scripps Institution of Oceanography, 2008; H.J. Walker, Jr., pers. comm. 2008).

California Academy of Sciences

55 Music Concourse Drive, Golden Gate Park, San Francisco, CA 94118
Specimens: 37.

The California Academy of Sciences in San Francisco, California, has in its catalogue, materials belonging to 37 specimens of great white shark. These materials are labeled with the following catalogue numbers: without catalogue number, CAS 65027, CAS 65031, CAS 82321, CAS 26366, CAS 26695, CAS 26793, CAS 26884, CAS 27158, CAS 48413, CAS 53045, CAS 55435, CAS 65029, CAS 65032, CAS 72090, CAS 79612, CAS 83397, CAS 214324, CAS 25844, CAS 26363, CAS 26781, CAS 27013, CAS 27014, CAS 27015, CAS 37917, CAS 26367, CAS 26694, CAS 26245, CAS 26308, CAS 17926, CAS 65022, CAS 26361, CAS 26376, CAS 26378, CAS 26680, CAS 55467, CAS 26678.

The item without catalogue number is a life-size replica of a pregnant female. This specimen measured 548.6 cm (18 ft) TL. This item was acquired by the California

A life-size replica of a 548.6 cm pregnant great white shark, currently on exhibit to the public at the Humboldt State University Natural History Museum in Arcata, California, on loan from the California Academy of Sciences (without cat. no.) (photograph courtesy Melissa L. Zielinski, Humboldt State University Natural History Museum, Arcata, California and California Academy of Sciences, San Francisco).

Academy of Sciences in 1966 and is currently on exhibit to the public at the Humboldt State University Natural History Museum, Arcata, California, on loan from the California Academy of Sciences (Melissa L. Zielinski, pers. comm. 2008).

The items with catalogue number CAS 65027 are teeth. These items were acquired by the California Academy of Sciences in 1966 (Ichthyology Collection Database of the California Academy of Sciences, 2008).

The items with catalogue number CAS 27158 are tooth fragments, which were preserved in liquid (75 percent ethyl alcohol). The tooth fragments were from a great white shark that attacked a skin diver on May 28, 1972, near Tomales Point, California. This item was acquired by the California Academy of Sciences in 1972. The collector was Helmuth Himmrich, and the determiner was Wilbur Irving Follett (Ichthyology Collection Database of the California Academy of Sciences, 2008). On May 28, 1972, about 1:30 P.M., a great white shark seized and seriously injured Mr. Helmuth Himmrich of Lodi, California, who was swimming in the channel between Bird Rock and the western shore of Tomales Point, Marin County, California (38°13'47"

N, 122°59'29" W). The water was between 3.6 and 4.6 m deep. Visibility was limited to 1.2 or 1.5 m as the water was murky with plankton. Himmrich had been diving for abalones. Mr. Himmrich's four companions were fishing nearby from a 548.6-cm boat. Four or five minutes before the attack, one of these anglers had caught a large greenling *Hexagrammos* sp. After returning to the surface, Himmrich was lying face down and was just starting to swim back to the boat when he was seized by the right thigh immediately below the buttock. Later, one of his companions, Charles Jones, said, "It was a shark. About 8 to 12 ft [243 to 365.8 cm] of him came out of the water." In Mr. Himmrich's own words, "He grabbed ahold of me, and then when I moved he really clamped down on me. He brought me up out of the water, except my head and chest, which were in the water. He shook me and then he released me. I never did see him. After he released me I swam, butterfly fashion, to where they could reach my arms and pull me into the boat." After his companions had bound his thigh with a leather belt, they took him to Dillon Beach, where he received further first aid. A Coast Guard helicopter then transported him to Crissy Field, San Francisco. From there he was taken by ambulance to Letterman General Hospital, where he was attended by John E. Hutton, Jr., M.D., Lieutenant Colonel, Medical Corps, U.S. Army. The most serious of Mr. Himmrich's wounds was a circumferential incision, about 9 inches long and 3 inches deep, immediately below the right buttock. A year later, some feeling returned to the injured leg, and some muscle movement was possible. From the deepest wound, Dr. Hutton removed three small tooth fragments (catalogue number CAS 27158) that are traceable to the white shark (Follett, 1974).

The items with catalogue number CAS 65031 are teeth. Almost surely these teeth are from the same great white shark that attacked Mr. Himmrich on May 28, 1972, near Tomales Point, California (in fact, they are labeled with same location and same collector) (Ichthyology Collection Database of the California Academy of Sciences, 2008). Strangely, Follett (1974), in his detailed account of this attack, did not report of any teeth of this white shark being labeled with catalogue number CAS 65031 at the California Academy of Sciences, only citing the previously mentioned tooth fragments labeled with catalogue number CAS 27158.

Catalogue no. CAS 82321 is a whole specimen, which was preserved in liquid (55 percent isopropyl alcohol) (location: Tank 15). This specimen measured 106.7 cm (3.5 ft) TL. This item was acquired by the California Academy of Sciences in 1995. The shark was received frozen from Steinhart Aquarium, the California Academy of Sciences Aquarium (Ichthyology Collection Database of the California Academy of Sciences, 2008).

Catalogue no. CAS 26366 was preserved in liquid (75 percent ethyl alcohol). The female great white shark was captured on July 29, 1959, in Tomales Bay, California. The specimen measured 277.4 cm TL and weighed 198.7 kg (438 lb). There were no contents in the stomach. Morphometric measurements of the whole specimen were taken by Wilbur Irving Follett and Lillian J. Dempster and are still available at the California Academy of Sciences. This item was acquired by the California Academy of

Sciences in 1959. The collectors were Felix Konatich and Tony Konatich (Wilbur Irving Follett, unpublished data; Ichthyology Collection Database of the California Academy of Sciences, 2008).

The items with catalogue number CAS 26695 were a set of jaws and vertebral column, which were preserved dry and in liquid, (ethyl alcohol) (location: Bone room/Annex oversize) respectively. The jaws and vertebral column were from a male great white shark that was captured on November 13, 1959, off Bolinas Bay, California. This specimen measured 267.3 cm TL and weighed 171.6 kg (378.25 lb). Morphometric measurements of the whole specimen were taken by Wilbur Irving Follett and Lillian J. Dempster and are still available at the California Academy of Sciences. This item was acquired by the California Academy of Sciences in 1959 and the collector was Philip Vella. An incorrect catalogue number (26392) was written on vertebral column segments. The centra 40 and 41 were loaned for destructive sampling (Wilbur Irving Follett, unpublished data; Ichthyology Collection Database of the California Academy of Sciences, 2008).

Catalogue no. CAS 26793 was preserved in liquid (75 percent ethyl alcohol). The female great white shark was captured on September 7, 1960, in Tomales Bay, California. This specimen measured 196.7 cm TL and weighed 69.8 kg (154 lb). There were no contents in the stomach. Morphometric measurements of the whole specimen were taken by Wilbur Irving Follett and J.D. Hopkirk and are still available at the California Academy of Sciences. This item was acquired by the California Academy of Sciences in 1960. The collector was Felix Konatich (Wilbur Irving Follett, unpublished data; Ichthyology Collection Database of the California Academy of Sciences, 2008).

The items with catalogue number CAS 26884 are teeth. The teeth were from a female great white shark that attacked a boat on September 10–12, 1959, between False Klamath and Klamath River mouths, California. This item was acquired by the California Academy of Sciences in 1960. The collector was Henry Tervo, and the determiner was Wilbur Irving Follett (Ichthyology Collection Database of the California Academy of Sciences, 2008; Follett, 1974). Henry Tervo, a commercial fisherman out of Eureka, California, reported an attack on his 11-m (36-ft) boat *Hunter*:

> I was salmon trolling on September 10–12, 1959, about 3:30 to 4 P.M., PST, about 4 miles offshore, between False Klamath and the Klamath River, Del Norte County, California, in about 30 to 35 fathoms. The sea was choppy and the water warm and turbid. The bottom was probably mud. I judged that I had hooked a "shaker" [undersized salmon] on my port line, and I was reaching down to release the clamp on the port side of the stern of the boat when I saw something coming at me like a torpedo. It was a shark, about 15 ft [457.2 cm] long. It grabbed the lower guard of the stern of the boat in its mouth. Its impetus raised its body and tail out of water to such a height that I thought it would topple into the cockpit. When it released itself it went around under the stern and came up on the starboard side. I went and got the gun and when I came back it was making a turn toward the boat again. Then it went down and I never saw it after that. Captain of Wardens Walt Gray and another warden took three teeth from the side of the boat while the boat was tied to the dock at Eureka. The photographer of the Richmond Independent

photographed the boat yesterday, but the scars had been largely obscured by the work done in dry dock [Follett, 1974].

Catalogue no. CAS 48413 is a set of jaws, which were preserved dry (location: Bone room). The jaws were from a great white shark that was captured around 1935 off Pigeon Point, California. This specimen was estimated to measure over 365.8 cm (12 ft) TL and to weigh around 453.6 kg (1000 lb). It was caught on a baited hook attached to a set line. This item was acquired by the California Academy of Sciences in 1960. The collector was Mr. Thompson (Ichthyology Collection Database of the California Academy of Sciences, 2008).

Catalogue no. CAS 53045 is a whole specimen, which was preserved in liquid (75 percent ethyl alcohol) (location: Tank 14). The male great white shark was captured on August 2, 1983, two miles due west of Ventura Marina entrance, California (34°14'N; 119°20'W). This specimen measured 130 cm SL and was caught in a monofilament gill net. The water temperature at the time of capture was 65°F. This item was acquired by the California Academy of Sciences in 1983. The collectors were commercial fisherman Ben Henke and John C. Hewitt, and the determiner was John C. Hewitt. This great white shark is the same specimen reported as being kept in captivity at the Steinhart Aquarium in August 1983 (see page 175, "Early '80s — August 1983: Steinhart Aquarium, California Academy of Sciences") (John C. Hewitt, pers. comm. 2009; Hewitt, 1984; Ichthyology Collection Database of the California Academy of Sciences, 2008).

Catalogue no. CAS 55435 is a whole specimen, which was preserved in liquid (75 percent ethyl alcohol) (location: Tank 11). This great white shark was captured on July 7, 1984, two miles due west of Ventura boat harbor entrance, California. It was caught in a monofilament floating gill net and this item was acquired by the California Academy of Sciences in 1984. The collector and determiner was John C. Hewitt. There is no doubt that this great white shark is the same 150 cm female specimen reported as being kept in captivity for three days at the Steinhart Aquarium in July 1984 after being captured two miles off the beach in a monofilament halibut gill net and towed into Ventura harbor (see page 176, "July 1984: Steinhart Aquarium, California Academy of Sciences") (Hewitt, 1984; Ichthyology Collection Database of the California Academy of Sciences, 2008).

The items with catalogue number CAS 65029 are teeth (Bone room). The teeth were from a great white shark that washed ashore at Camp Pendleton near San Clemente, California. This specimen measured 167.6 cm (5.5 ft) TL and weighed 68.0 kg (150 lb). The collector was B. Vernarecci, and the determiner was Wilbur Irving Follett (Ichthyology Collection Database of the California Academy of Sciences, 2008).

The items with catalogue number CAS 65032 are teeth. The teeth were from a female great white shark that was captured on June 30, 1942, off Catalina Island, California. This specimen measured 166 cm TL and weighed 39.5 kg (87 lb). This is the same specimen reported in Roedel and Ripley (1950) (Bigelow and Schroeder, 1948; Ichthyology Collection Database of the California Academy of Sciences, 2008).

Catalogue no. CAS 72090 is a set of jaws. The jaws were from a male great white shark that was captured on July 31, 1959, off Selva Beach, California. This specimen measured 368.3 cm (12 ft 1 in) TL and weighed 380.1 kg (838 lb) (Ichthyology Collection Database of the California Academy of Sciences, 2008).

The items with catalogue number CAS 79612 are teeth, four from the lower jaw and eight from the upper jaw. The teeth were from a great white shark that was captured on October 26, 1943, off Bodega Bay, California. This specimen measured approximately 701 cm (23 ft) TL. The collector was Standard Fisheries of San Francisco. Three teeth of this specimen were illustrated by Keith W. Cox and appear in Roedel and Ripley (1950) (Ichthyology Collection Database of the California Academy of Sciences, 2008; Roedel and Ripley, 1950).

The items with catalogue number CAS 83397 are teeth, which were preserved dry. The teeth were from a great white shark that was captured on November 12, 1943, off Fort Bragg, California (Ichthyology Collection Database of the California Academy of Sciences, 2008).

Catalogue no. CAS 214324 is a vertebral column, which was preserved in liquid (75 percent ethyl alcohol). The vertebral column was from a great white shark that was captured on August 23, 1984, 10 miles south of Santa Cruz Island, California. This specimen measured 533.4 cm TL and weighed 1474 kg. The collector was Leonard J.V. Compagno (Ichthyology Collection Database of the California Academy of Sciences, 2008).

The items with catalogue number CAS 25844 are 90 vertebrae, 75 upper teeth and 93 lower teeth, which were preserved dry. The vertebrae and teeth were from a female great white shark that was captured on September 9, 1936, approximately 0.25

A 167.6 cm TL female white shark captured on September 9, 1936, off Malibu, California, of which 90 vertebrae, 75 upper teeth and 93 lower teeth are preserved at the California Academy of Sciences in San Francisco, California with cat. no. CAS 25844 (photograph courtesy W.I. Follett, Dept. of Ichthyology, California Academy of Sciences).

mi. southeast of Malibu, California (34°2'0"N; 118°41'0"W). This specimen measured 167.6 cm (5.5 ft) TL and weighed 42.2 kg (93 lb). It was caught on hook and line using a whole mackerel as bait. This item was acquired by the California Academy of Sciences in 1951 (Ichthyology Collection Database of the California Academy of Sciences, 2008; Follett, 1966).

Catalogue no. CAS 26363 was preserved in liquid (75 percent ethyl alcohol). The male great white shark that was captured on July 25, 1959, off Indian Beach, California. This specimen measured 314.3 cm TL and weighed 283.5 kg (625 lb). It was caught in a net that measured 450' × 25' with 7" mesh. The depth of capture was 5.5 m. This shark had the remains of a bat eagle ray *Myliobatis californica* (Gill, 1865) in its stomach (also preserved in the California Academy of Sciences in San Francisco with catalogue no. CAS 26364). Morphometric measurements of the whole specimen were taken by Wilbur Irving Follett and Lillian J. Dempster and are still available at the California Academy of Sciences. This item was acquired by the California Academy of Sciences in 1959. The collectors were Felix Konatich and Tony Konatich (Wilbur Irving Follett, unpublished data; Ichthyology Collection Database of the California Academy of Sciences, 2008; Follett, 1974).

The items with catalogue number CAS 26781 are a toothless set of jaws, separate

A 295.9 cm TL female great white shark that was captured on July 2, 1960, in Tomales Bay, southeast of Hog Island. The toothless set of jaws, separate teeth, and vertebral column are preserved at the California Academy of Sciences in San Francisco, California with cat. no. CAS 26781 (photograph courtesy Wilbur Irving Follett, Dept. of Ichthyology, California Academy of Sciences).

teeth, and vertebral column. The jaws and teeth were preserved dry, and the vertebral column was preserved in liquid (alcohol). The jaws, teeth and vertebral column were from a female great white shark that was captured on July 2, 1960, in Tomales Bay, 300 yds. southeast of Hog Island, California. The depth of capture was 6.1 m. This specimen measured 295.9 cm TL and weighed 190.1 kg (419 lb). It was caught on a 3 inch hook baited with a salmon head. This shark had a tope shark *Galeorhinus galeus* (also preserved in the California Academy of Sciences in San Francisco with catalogue no. CAS 26782) and a bat eagle ray *Myliobatis californica* (also preserved in the California Academy of Sciences in San Francisco with catalogue no. CAS 26783) in its stomach. Morphometric measurements of the whole specimen were taken by Wilbur Irving Follett and Lillian J. Dempster and are still available at the California Academy of Sciences. This item was acquired by the California Academy of Sciences in 1960. The collectors were Frank Spenger, Jr., and Richard Richarz (vessel *Marcella II*) (Wilbur Irving Follett, unpublished data; Ichthyology Collection Database of the California Academy of Sciences, 2008).

Catalogue no. CAS 27013 is a skin patch. The skin patch was from a female great white shark that was captured on December 27, 1960, off Stinson Beach, California. The depth of capture was approximately 11 fathoms. This specimen measured 241.9 cm (7 ft 11.25 in) TL. The stomach contained Chinook salmon *Oncorhynchus tshawytscha*. Morphometric measurements of the whole specimen were taken by Wilbur Irving Follett and Lillian J. Dempster and are still available at the California Academy of Sciences. It was acquired by the California Academy of Sciences in 1961. The collectors were C.J. Crivello and J. Crivello (Wilbur Irving Follett, unpublished data; Ichthyology Collection Database of the California Academy of Sciences, 2008).

The items with catalogue number CAS 27014 are a set of jaws and a vertebral column, which were preserved dry and in liquid (ethyl alcohol), respectively. The jaws were from a female great white shark that was captured on January 2, 1961, off Stinson Beach, California. The depth of capture was 11 fathoms. This specimen measured 327.0 cm TL and weighed 326.6 kg (720 lb). It was caught in seabass gill net (400 fathoms long with 6.5 in mesh). This shark had a quadrate from a striped bass *Morone saxatilis* (also preserved in the California Academy of Sciences in San Francisco with catalogue no. CAS 200720) and a bat eagle ray *Myliobatis californica* in its stomach. Morphometric measurements of the whole specimen were taken by Wilbur Irving Follett and Lillian J. Dempster and are still available at the California Academy of Sciences. This item was acquired by the California Academy of Sciences in 1961. The collectors were Carlo and Joe Crivello, and the determiners were Wilbur Irving Follett and Lillian J. Dempster (Wilbur Irving Follett, unpublished data; Ichthyology Collection Database of the California Academy of Sciences, 2008).

The items with catalogue number CAS 27015 are a set of jaws, skin patch, vertebral column, pelvic fins and claspers. The jaws were preserved dry, the vertebral column was preserved in liquid (ethyl alcohol), and the pelvics and claspers were preserved in liquid (formalin). The jaws, skin patch, vertebral column, pelvics and claspers were from a male great white shark that was captured on December 28, 1960, off Stinson Beach, California. This specimen measured 265.0 cm TL and weighed 180.5 kg (398 lb). The stomach contained a Pacific rock crab *Cancer antennarius*, a brown smooth-hound *Mustelus henlei*, three Chinook salmons *Oncorhynchus tshawytscha*, a black rockfish *Sebastes melanops*, and a lingcod *Ophiodon elongatus*. Morphometric measurements of the whole specimen were taken by Wilbur Irving Follett and Lillian J. Dempster and are still available at the California Academy of Sciences. This item was acquired by the California Academy of Sciences in 1959. The collectors were Carlo and Joe Crivello, and the determiners were Wilbur Irving Follett and Lillian J. Dempster. The set of jaws is on "permanent" display in "Life Through Time" (Wilbur Irving Follett, unpublished data; Ichthyology Collection Database of the California Academy of Sciences, 2008).

The items with catalogue number CAS 37917 are a gill arch, vertebrae, and other unspecified parts, which were preserved in liquid (75 percent ethyl alcohol). The gill arch, vertebrae, and other parts were from a great white shark that was captured on October 9, 1976, just outside the entrance to Tomales Bay, California. It was caught in a halibut gill net. This item was acquired by the California Academy of Sciences in 1976 and the collector was Pete Halley. There is no doubt that this great white shark is the same 213.4 cm (7 ft) specimen reported as being kept in captivity for less than one day at the Steinhart Aquarium in 1976 after being captured just outside the entrance of Tomales Bay, California, by halibut gillnetters Pete Halley and his brother (see page 171, "October 1976: Steinhart Aquarium, California Academy of Sciences, San Francisco, California") (Ichthyology Collection Database of the California Academy of Sciences, 2008; Powell, 2001; David C. Powell, pers. comm. 2008, 2009).

Catalogue no. CAS 26367 was from a male great white shark that was captured on July 30, 1959, off Young Landing, California. This specimen measured 218.4 cm TL and weighed 98.9 kg (218 lb). Morphometric measurements of the whole specimen were taken by Wilbur Irving Follett and Lillian J. Dempster and are still available at the California Academy of Sciences. This item was acquired by the California Academy of Sciences in 1959. The collectors were Felix Konatich and Tony Konatich (Wilbur Irving Follett, unpublished data; Ichthyology Collection Database of the California Academy of Sciences, 2008).

Catalogue no. CAS 26694 is an almost whole specimen (centra 56 and 57 were loaned for destructive sampling and the jaws, strip of skin, and vertebral column were removed), which was preserved dry and in liquid (ethyl alcohol). The male great white shark was captured on November 9, 1959, off Stinson Beach, California. This specimen measured 225.4 cm TL and weighed 101.6 kg (224 lb). Morphometric measurements of the whole specimen were taken by Wilbur Irving Follett and Lillian J. Dempster and are still available at the California Academy of Sciences. This item was acquired by the California Academy of Sciences in 1959 and the collector was Philip Vella aboard the vessel *Chief* (Wilbur Irving Follett, unpublished data; Ichthyology Collection Database of the California Academy of Sciences, 2008).

The items with catalogue number CAS 26245 are dermal denticles, jaws and teeth. The dermal denticles, jaws and teeth were from a great white shark that was captured on January 23, 1957, 0.25 miles off mouth of the Salinas River in California. The depth of water was 15 fathoms. This specimen measured 510.5 cm (16 ft 9 in) TL and was harpooned. Morphometric measurements of the whole specimen were taken by Wilbur Irving Follett and Jack Daniels and are still available at the California Academy of Sciences. This item was acquired by the California Academy of Sciences in 1957. The collectors were W.H. Tomlinson and Jack Daniels aboard the vessel *White Angel*, and the determiner was Wilbur Irving Follett (Wilbur Irving Follett, unpublished data; Ichthyology Collection Database of the California Academy of Sciences, 2008; Follett, 1966).

Catalogue no. CAS 26308 is a tooth fragment. The tooth fragment was removed from a sea otter found dead on the shore on February 6, 1958, at Pebble Beach, California. This item was acquired by the California Academy of Sciences in 1958. The collector was H.V. Shelby, and the determiners were Wilbur Irving Follett and Leo S. Dempski. This was the first verified white shark tooth fragment from a lacerated sea otter *Enhydra lutris* (Ichthyology Collection Database of the California Academy of Sciences, 2008; Orr, 1959).

Catalogue no. CAS 17926 is an almost whole specimen (teeth sets extracted), which was preserved in liquid (75 percent ethyl alcohol). The male great white shark was captured on July 14, 1945, between Scripps Canyon and Mussel Rock off Torrey Pines Beach in San Diego, California (32°53.3'N, 117°16.0'W). This specimen measured 137 cm TL (117 cm FL). The depth of water was 27.4 m. It was caught with setline by an Encinitas fisherman. This specimen was acquired by the California Academy of Sciences from the Scripps Institution of Oceanography (where it was labeled with catalogue number SIO 45-108). The collector and determiner was Carl Leavitt Hubbs (H.J. Walker, Jr., pers. comm. 2008; Ichthyology Collection Database of the California Academy of Sciences, 2008; Ichthyology Collection Database of the Scripps Institution of Oceanography, 2008).

The items with catalogue number CAS 65022 are vertebrae, which were preserved dry. The vertebrae were from a female great white shark that was captured on November 15, 1945, off La Jolla, just north of Scripps Pier, California. The depth of capture was 9.1 m and this specimen measured approximately 182.9 cm (6 ft) TL. The collector was G. Schillriff, and the determiner was Carl Leavitt Hubbs (Ichthyology Collection Database of the California Academy of Sciences, 2008).

The items with catalogue number CAS 26361 are a set of jaws and a vertebral column. The jaws and vertebral column were from a female great white shark that was captured on July 12, 1959, 1.5 miles off Goleta Point, California, in 27.4 m of water. This specimen measured 249.2 cm TL and weighed 120.2 kg (265 lb). It was caught along a sand bottom with a 7" stretched nylon mesh gill net set by Lawrence Castagnola from his boat *Americo* when he pulled the net at 6:30 A.M. Morphometric measurements of the whole specimen were taken by Wilbur Irving Follett and Lillian J.

Dempster and are still available at the California Academy of Sciences. This item was acquired by the California Academy of Sciences in 1959. The collector was L. Castagnola (Wilbur Irving Follett, unpublished data; Ichthyology Collection Database of the California Academy of Sciences, 2008; Dave Catania, pers. comm. 2008; Follett, 1966).

The items with catalogue number CAS 26376 are a set of jaws and vertebrae. The jaws and vertebrae were from a male great white shark that was captured on August 10, 1959, off Soquel Point, California. This specimen measured 194.3 cm TL and weighed 61.7 kg (136 lb). Morphometric measurements of the whole specimen were taken by Wilbur Irving Follett and Lillian J. Dempster and are still available at the California Academy of Sciences. This item was acquired by the California Academy of Sciences in 1959. The collector was A. Carniglia (Wilbur Irving Follett, unpublished data; Ichthyology Collection Database of the California Academy of Sciences, 2008).

Catalogue number CAS 26378 is a set of jaws. The jaws were from a male great white shark that was captured on August 28, 1959, off Palm Beach, California. This specimen measured 195.9 cm (6 ft 5⅛ in) TL and weighed 54.4 kg (120 lb). This item was acquired by the California Academy of Sciences in 1959. The collector was A. Carniglia (Ichthyology Collection Database of the California Academy of Sciences, 2008).

The items with catalogue number CAS 26680 are a set of jaws and vertebrae. The jaws and vertebrae were from a male great white shark that was captured on August 7, 1959, off Capitola, California. The depth of capture was 13–14 fathoms. This specimen measured 192.5 cm TL and weighed 65.8 kg (145 lb). It was caught in a drift net that was 300' × 17' with 7" mesh. Morphometric measurements of the whole specimen were taken by Wilbur Irving Follett and Manuel Silva and are still available at the Califor-

A 192.5 cm TL male great white shark that was captured on August 7, 1959, off Capitola, California. The set of jaws and vertebrae are preserved at the California Academy of Sciences in San Francisco, California, with cat. no. CAS 26680 (photograph courtesy Wilbur Irving Follett, Dept. of Ichthyology, California Academy of Sciences).

nia Academy of Sciences. This item was acquired by the California Academy of Sciences in 1959. The collector was V. Cardinalli aboard the vessel *Little Rose* (Wilbur Irving Follett, unpublished data; Ichthyology Collection Database of the California Academy of Sciences, 2008; Follett, 1966).

Catalogue no. CAS 55467 is a set of jaws. The jaws were from a great white shark that was captured off Santa Cruz, California. This item was acquired by the California Academy of Sciences in 1960. The collector was an unknown fisherman and was received by John Strobeen, around 1930 (Ichthyology Collection Database of the California Academy of Sciences, 2008).

The items with catalogue no. CAS 26678 are a vertebral column (centra 98 and 99 were loaned for destructive sampling) and a tooth. The vertebral column was preserved in liquid (isopropyl alcohol) and the tooth was dried. The vertebral column and tooth were from a male great white shark that was captured on July 31, 1959, off La Selva Beach, California. This specimen measured 368.3 cm TL and weighed 380.1 kg (838 lb). Morphometric measurements of the whole specimen were taken by Wilbur Irving Follett and are still available at the California Academy of Sciences. This item was acquired by the California Academy of Sciences in 1959. The collector was J.L. Olivieri aboard the vessel *Maria*, and the determiner was Wilbur Irving Follett (Wilbur Irving Follett, unpublished data; Ichthyology Collection Database of the California Academy of Sciences, 2008).

CONNECTICUT

Department of Marine Sciences, University of Connecticut

1080 Shennecossett Road, Groton, CT 06340
Specimens: 1.

The Department of Marine Sciences of the University of Connecticut in Groton, Connecticut, has in its catalogue materials belonging to one specimen of great white shark. This material is without catalogue number.

The item without catalogue number is a great white shark replica that was made from a cast of the original specimen. This male great white shark was caught 8 to 10 miles east of Block Island, Connecticut, in 34.4 m of water, on August 5, 1983, by Gregory Dubrule and Ernest Celotto, professional deep sea fishermen. After harpooning, a five-hour struggle followed and the shark was dispatched with two loads of buckshot. The shark was feeding on a 12.2-m (40-ft) dead whale carcass. This great white shark measured 484 cm TL and weighed 1260.5 kg (2779 lb). Its liver weighed 205.5 kg (453 lb) and its heart weighed 2.3 kg (5 lb 2 ounces). There was an outraged public outcry at the sport fishermen who caught the shark, and they chose to donate it to the Connecticut State Museum of Natural History to mitigate the bad publicity. This item was

A cast replica of a 484 cm TL male great white shark caught east of Block Island, Connecticut, on August 5, 1983. This replica is on exhibit to the public, hanging in a cafeteria at the entrance to one of the buildings of the Department of Marine Sciences of the University of Connecticut in Groton, Connecticut, without cat. no. (photograph courtesy J. Evan Ward, Department of Marine Sciences of the University of Connecticut, Groton).

acquired by the Connecticut State Museum of Natural History on August 7, 1983. A fiberglass epoxy resin mold was manufactured at the Mystic Marinelife Aquarium by a Long Island firm. The final taxidermy and painting was done by Peter Wilson of Taxidermy Plus in Hamden, Connecticut. The shark reproduction measures 457.2 cm TLn and weighs 90.7 kg (200 lb). The upper jaw perimeter is approximately 82.3 cm, and the teeth measure 3.9 cm UAE1. The brain, heart, liver, and ring valve, as well as samples of other organs were saved for study by scientists in several states. Scrapings were made from the teeth and analyzed by Dr. John Buck of the University of Connecticut, Avery Point. Thousands of parasites were removed from the inside and the outside of the shark, and were studied by George Benz, a former University of Connecticut graduate student and authority on shark parasites. The teeth turned out to be somewhat problematic because the company that the museum paid to make the original mold, despite the museum's wishes, actually prepared two casts of the shark and used some of the teeth in that second cast as well as the first. Several years ago, shortly after 2001, the Connecticut State Museum of Natural History sent the cast to the Avery Point Campus of the University, to the Department of Marine Sciences. The item is currently on exhibit to the public, hanging in a cafeteria at the entrance to one of the

Department of Marine Science buildings (Cheri F. Collins, pers. comm. 2008; Susan Hochgraf, pers. comm. 2008; Evan Ward, pers. comm. 2008).

FLORIDA

Florida Museum of Natural History, University of Florida

University of Florida Cultural Plaza, SW 34th Street and Hull Road, Gainesville, FL 32611-2710
Specimens: 13.

The Florida Museum of Natural History in Gainesville, Florida, has in its catalogue materials belonging to 13 specimens of great white shark. These materials are labeled with the following catalogue numbers: UF 42630, UF 44771, UF 47943, UF 48285, UF 48814, UF 98391, UF 98392, UF 103732, UF 111019, UF 111026, UF 111027, UF 111028, UF 162499.

The items with catalogue number UF 42630 are a first dorsal fin and dermal denticles, which were preserved in liquid (alcohol). The dorsal fin and dermal denticles were from a great white shark that was captured 18 mi off Daytona Beach, Florida. This item was cataloged in 1985. This item is not currently on exhibit to the public (Ichthyology Collection Database of the University of Florida, 2008; Robert H. Robins, pers. comm. 2008).

The items with catalogue number UF 44771 are dermal denticles, a vertebra centrum and a clasper, which were preserved in liquid (alcohol) (location: Jar/White Tank 2). The dermal denticles, vertebra centrum and clasper were from a male great white shark that was captured on August 6, 1986, 20 mi SE of Montauk Point, New York. This great white shark measured a little over 518.2 cm (17 ft) TL and weighed 1554.5 kg (3,427 lb). It was caught by angler Don Braddick with the help of Frank Mundus on the boat *The Crickett II*. This white shark is not only a New York State marine fishing record but also the largest fish caught on rod and reel. This item was cataloged in 1986. The item is not currently on exhibit to the public. This is how Frank Mundus described the capture of this huge white shark: "This shark was a little over 17 ft [518.2 cm] long and took us an hour and 40 minutes to put the first gaff into it and then, after struggling for another hour, we finally got the fish secured with a tail rope, and dragged it home behind the boat. Back at the dock at midnight, it took us another three or four hours to weigh in the fish, take photos and put him on a bed of ice for the night at Montauk Marine Basin." The great white shark was examined by National Marine Fisheries Service scientists Jack Casey, Harold L. "Wes" Pratt and their crew from the National Marine Fisheries Service Narragansett Laboratory, Rhode Island. A life-sized fiberglass replica of this specimen is on display hanging at the docks of the Star Island Marina in

Montauk Harbor (DeMarte, 2007; Ichthyology Collection Database of the University of Florida, 2008; Mundus, 2000; Robert H. Robins, pers. comm. 2008).

Catalogue no. UF 47943 (previous catalogue number: Z-4672) is a partial vertebral column, which was preserved dry. The vertebral column was from a great white shark that was captured off Daytona Beach, Florida. This item was cataloged in 1986. This item is not currently on exhibit to the public (Ichthyology Collection Database of the University of Florida, 2008; Robert H. Robins, pers. comm. 2008).

Catalogue no. UF 48285 (previous catalogue number: Z1482) is a set of jaws, which was preserved dry. The jaws were from a great white shark that was captured on March 11, 1959. This great white shark measured 261.6 cm (103 in) TL and weighed 176.9 kg (390 lb). This item was cataloged in 1990. This item is not currently on exhibit to the public (Ichthyology Collection Database of the University of Florida, 2008; Irv Quitmyer, pers. comm. 2008; Robert H. Robins, pers. comm. 2008).

Catalogue no. UF 48814 (previous catalogue number: Z7248) is a partial vertebral column, which was preserved dry. The vertebral column was from a great white shark that was captured on January 4, 1994, off Georgia. This item was cataloged in 1995 and is not currently on exhibit to the public (Ichthyology Collection Database of the University of Florida, 2008; Robert H. Robins, pers. comm. 2008).

The items with catalogue number UF 98391 are a head and upper lobe of the caudal fin, which were preserved in liquid (alcohol) (location: White Tank 7). The head and upper lobe of the caudal fin were from a great white shark that was captured on January 4, 1995, approximately 10 miles east of Daytona Beach, Florida (GPS, from 29°10'25"N, 80°34'32"W, to 29°29'8"N, 80°40'35"W). The shark was captured between 7:30 and 11:55 A.M. It was caught on a longline set at a depth of 24.3–27.1 m. This specimen measured 194 cm TL and 183 cm FOR. The collector and determiner was Craig J. Plizga on fishing vessel *Marsea*. This item was cataloged in 1995 and is not currently on exhibit to the public (Ichthyology Collection Database of the University of Florida, 2008; Robert H. Robins, pers. comm. 2008).

Catalogue no. UF 98392 is a heart, which was preserved in liquid (alcohol). The female great white shark was captured on February 25, 1995, approximately 10 miles east of Cape Canaveral off Florida. It was caught between 10:32 A.M. and 12:40 (GPS, from 28°40'23"N, 80°20'7"W, to 28°43'44"N, 80°24'39"W). It was caught on a longline at a depth of 21.3–24.1 m. This specimen measured 279 cm TL and 248 cm FOR and its round weight was 229.5 kg (506 lb). The collector and determiner was Craig J. Plizga on fishing vessel *Ms Debb*. This item was cataloged in 1995 and is not currently on exhibit to the public (Ichthyology Collection Database of the University of Florida, 2008; Robert H. Robins, pers. comm. 2008).

The items with catalogue number UF 103732 are a whole specimen and a DNA tissue sample immersed in Longimire's DNA Buffer, which were preserved in liquid (formalin). The great white shark was captured on January 6, 1996, west of Wrightsville Beach, North Carolina (GPS, 34°19'35"N, 76°26'5"W, to 34°5'8"N, 76°37'9"W).

The surface temperature was 19°C. It was caught on a longline set at a depth of 26.5–39.1 m. The collector and determiner was Chris Jensen aboard the vessel *Tarbaby*. This item was cataloged in 1997 and is not currently on exhibit to the public (Ichthyology Collection Database of the University of Florida, 2008; Robert H. Robins, pers. comm. 2008).

Catalogue no. UF 111019 is a whole specimen, which was preserved in liquid (alcohol). The male great white shark was captured on February 26, 1998, approximately 43 miles ESE of Daytona Beach, Florida (GPS, 29°9'15,6"N, 80°17'36"W, to 29°7'32,4"N, 80°23'37,8"W). The water temperature was 64.7–71.3° F and it was caught between 8:25 A.M. and 11:04 A.M. It was caught on a longline set at a depth of 42.7–29.9 m. This specimen measured 178 cm TL and 163 cm FOR. The collectors were Joe Ludwig, Brett Robbins, Kevin M. Johns, and the determiner was Kevin M. Johns aboard the fishing vessel *Sea Dancer*. This item was cataloged in 1999. This specimen was kept whole for a potential future public display. However, this item is not currently on exhibit to the public (Ichthyology Collection Database of the University of Florida, 2008; Kevin M. Johns, pers. comm. 2008; Robert H. Robins, pers. comm. 2008).

The items with catalogue numbers UF 111026 and UF 111027 are eyes and a vertebral sample, which were preserved in liquid (vertebral sample in ethanol, never fixed in formalin, and eyes in isopropyl alcohol, originally fixed in formalin). The eyes and vertebrae were from a female great white shark that was captured on January 26–27, 1999, off Daytona Beach, Florida (GPS, from 29°43'32,4"N, 81°0'46,8"W, to 29°43'30"N, 80°54'3,6"W). It was caught between 8:06 P.M. and 1:33 P.M. The surface water temperature was 21.35°C. It was caught on a longline set at a depth of 19.8–26.2 m. This specimen measured 273 cm TL and 248 cm FOR. A full body dissection was executed with external and internal measurements recorded. The collector and determiner was Kevin M. Johns aboard the fishing vessel *Sea Dancer*. It is unclear why these items, that are from a single white shark, were labeled with two different catalogue numbers. These items were cataloged in 1999 and are not currently on exhibit to the public (Ichthyology Collection Database of the University of Florida, 2008; Kevin M. Johns, pers. comm. 2008; Robert H. Robins, pers. comm. 2008).

The items with catalogue number UF 111028 are heart, gonads, eyes, and dermal denticle samples cut from below first dorsal fin, laterally above the pelvis, and above the last gill slit, which were preserved in liquid (isopropyl alcohol). The heart, gonads, eyes, and dermal denticles were from a male great white shark that was captured on February 4–5, 1999, off Daytona Beach, Florida (GPS, from 29°51'30,6"N, 81°2'36"W, to 29°52'24,6"N, 81°56'7,2"W). The surface temperature was 20.5°C and it was caught between 20:00 and 12:00. It was caught on a longline set at a depth of 21–23.8 m. This specimen measured 227 cm TL and 207 cm FOR. A full body dissection was executed with external and internal measurements recorded. The collector and determiner was Kevin M. Johns aboard the fishing vessel *Sea Dancer*. This item was cataloged in 1999 and is not currently on exhibit to the public (Ichthyology Collection Database of the University of Florida, 2008; Kevin M. Johns, pers. comm. 2008; Robert H. Robins, pers. comm. 2008).

The items with catalogue number UF 162499 are three whole specimens, which were preserved in liquid (alcohol). The 3 great white sharks were captured more than 3 miles offshore of Huntington Beach, California. The depth of capture was 45.7 m (150 ft) and they were caught in a gillnet. These specimens measured 139.7, 154.9, and 160.0 cm (55, 61, and 63 in) TL. The collector was a commercial fisherman, Mike Sargeant. This item was cataloged in 2006 and it is not currently on exhibit to the public (Ichthyology Collection Database of the University of Florida, 2008; Robert H. Robins, pers. comm. 2008).

Mote Marine Laboratory
1600 Ken Thompson Parkway, Sarasota, FL 34236
Specimens: 4.

The Mote Marine Laboratory in Sarasota, Florida, has in its catalogue materials belonging to 4 specimens of great white shark. These materials are labeled with the following catalogue numbers: without catalogue number, GWS21, GWS25, GWS26.

The item without catalogue number is a set of jaws, which was preserved dry. The jaws were from a great white shark that was captured on February 19, 1967, 3.5 miles west of Midnight Pass, Sarasota, Florida. This specimen measured 337.8 cm (11 ft 1 in) TL and weighed 480.8 kg (1060 lb). The item is currently on exhibit to the public in the Mote Marine Laboratory Aquarium (Eugenie Clark, pers. comm. 2009; Robert E. Hueter, pers. comm. 2009; Debra A. Ingrao, pers. comm. 2009).

Catalogue no. GWS26 is a whole specimen, which was preserved by embalming with a formalin-based solution. The female great white shark was captured in June 2004, off the southern coast of California. This specimen measured approximately 274 cm TL and weighs 209 kg (this is preserved weight, not fresh weight). It was captured incidentally by a commercial halibut fisherman using a gill net. The skin color is unnatural, which was caused by one of the chemicals tested during preservation to determine the best chemicals to use for embalming very large specimens. This item was acquired by the Mote Marine Laboratory from a taxidermist as a donation some years ago. The collector asked to remain anonymous. The item is not currently on exhibit to the public (José I. Castro, pers. comm. 2009; Robert E. Hueter, pers. comm. 2009; Debra A. Ingrao, pers. comm. 2009).

Catalogue no. GWS25 is a whole specimen, which was embalmed with a formalin-based solution. The male great white shark was captured probably in 2004, off the southern coast of California. The depth of capture was 45.7 m. This specimen measured approximately 176 cm TL and weighs 69 kg (this is preserved weight, not fresh weight). It was captured incidentally by a commercial halibut fisherman using a gill net. The unnatural skin color was caused by one of the chemicals tested during preservation to determine the best chemicals to use for embalming very large specimens.

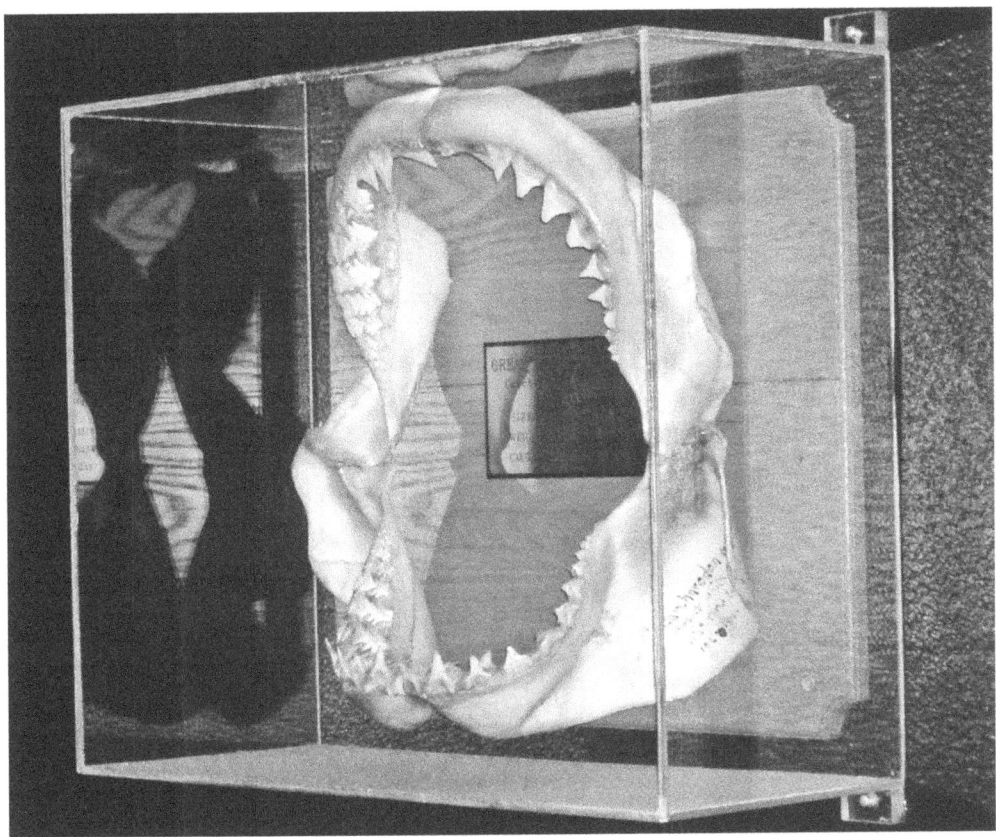

A set of jaws from a 337.8 cm TL great white shark captured on February 19, 1967, west of Midnight Pass, Sarasota, Florida. The jaws are on exhibit to the public in the Mote Marine Laboratory Aquarium in Sarasota, Florida (photograph courtesy Mote Marine Laboratory and aquarium, photograph by Debra A. Ingrao).

This item was acquired by the Mote Marine Laboratory from a taxidermist as a donation some years ago and the collector asked to remain anonymous. The item is currently on exhibit to the public in the Mote Marine Laboratory Aquarium (José I. Castro, pers. comm. 2009; Robert E. Hueter, pers. comm. 2009; Debra A. Ingrao, pers. comm. 2009).

Catalogue no. GWS21 is another whole specimen, which was embalmed with a formalin-based solution. The female great white shark was captured probably in 2004, off the southern coast of California. This specimen measured approximately 140 cm TL and weighs 22 kg (this is preserved weight, not fresh weight). It was captured incidentally by a commercial halibut fisherman using a gill net. Like the other specimens, the unnatural skin color was caused by one of the chemicals tested during preservation to determine the best chemicals to use for embalming very large specimens. This item was acquired by the Mote Marine Laboratory from a taxidermist as a donation some

Two great white sharks, a 176 cm TL male (cat. no. GWS25) and a 140 cm TL female (cat. no. GWS21), captured off the southern coast of California, embalmed with a formalin-based solution, on exhibit to the public in the Mote Marine Laboratory Aquarium in Sarasota, Florida (photograph courtesy Mote Marine Laboratory and Aquarium, photograph by Debra A. Ingrao).

years ago. The collector asked to remain anonymous. The item is currently on exhibit to the public in the Mote Marine Laboratory Aquarium (Robert E. Hueter, pers. comm. 2009; Debra A. Ingrao, pers. comm. 2009).

ILLINOIS

The Field Museum

1400 S. Lake Shore Dr, Chicago, IL 60605-2496
Specimens: 1.

The Field Museum in Chicago, Illinois, has in its catalogue materials belonging to one specimen of great white shark. These materials are labeled with the following catalogue number: FMNH 38335.

The items with catalogue number FMNH 38335 are a head (without teeth), left pectoral fin, right pectoral fin, first dorsal fin, pelvic fins, and caudal fin of a single white shark, which were preserved in liquid (70 percent ethyl alcohol). The head and fins were from a great white shark that was captured in April 1941, off Southern Florida by

3 — *Materials in Museums* — Illinois

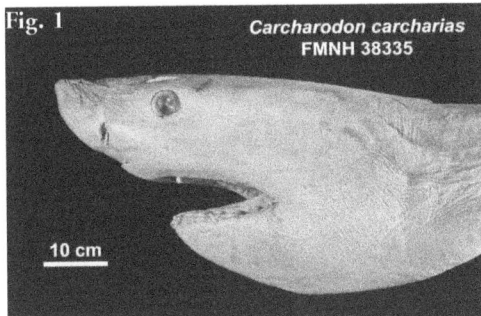
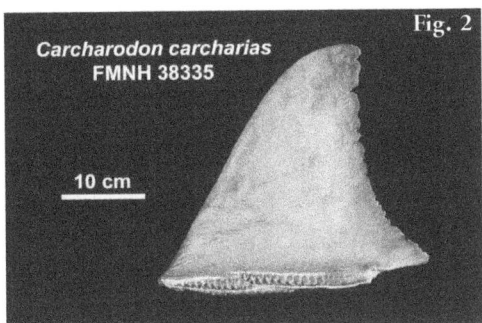
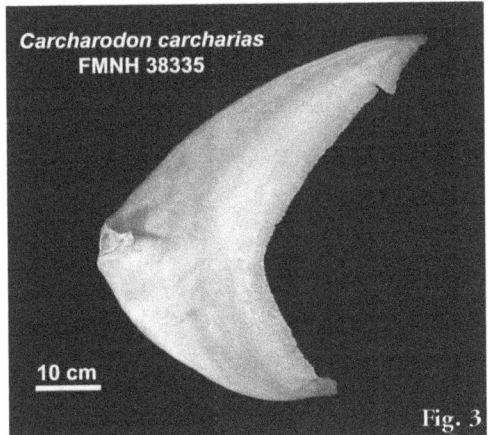
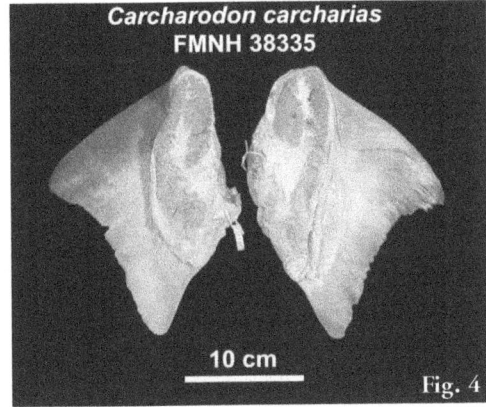
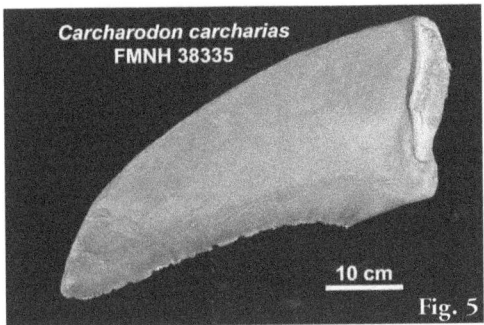
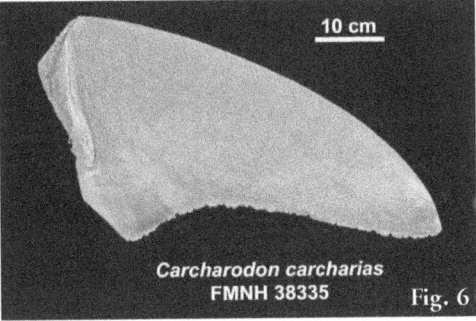

Parts from a great white shark that was captured in April 1941, off southern Florida, and preserved at the Field Museum in Chicago, Illinois, with cat. no. FMNH 38335. Figure 1 — head (without teeth). Figure 2 — First dorsal fin. Figure 3 — Caudal fin. Figure 4 — Pelvic fins. Figure 5 — Left pectoral fin. Figure 6 — Right pectoral fin (courtesy Philip Willink, Field Museum of Natural History, Chicago).

Florida Marine Products Inc. The DUJP is approximately 48.0 cm (Philip Willink, pers. comm. 2008; Mary Anne Rogers, pers. comm. 2007)

MASSACHUSETTS

Museum of Science
Science Park, Boston, MA 02114
Specimens: 1 (missing).

The Museum of Science in Boston, Massachusetts, held materials belonging to a specimen of great white shark. This material was probably without catalogue number. The item without catalogue number was a mounted specimen about 182.9 cm (six ft) long from Woods Hole, Massachusetts. Many years ago, at least until the late 1940s, the item was on exhibit to the public in the New England Museum of Natural History

A mounted white shark about 182.9 cm long from Woods Hole, Massachusetts, can be seen on the right (close to a swordfish *Xiphias gladius*) in this photo that shows the Main Hall of the New England Museum of Natural History (formerly the Boston Museum of Science), where it was on display until the late 1940s. A basking shark *Cetorhinus maximus* is in the center of the picture (courtesy Museum of Science, Boston).

(the previous name of the Boston Museum of Science), as is shown by an old black and white photograph and a short citation in Bigelow and Schroeder (1948). This specimen is no longer in the Museum of Science collection. At the Museum today are about 200 cards that once represented the ichthyology catalogue, including the card for the mounted great white shark, but there is no note as to its whereabouts. Most of the Museum of Science mounts were transferred to the Museum of Comparative Zoology at Harvard University in Cambridge, Massachusetts, and it is thought that the mounted great white shark followed the same destiny. However, there are no complete great white shark mounts preserved at Harvard's Museum of Comparative Zoology today. It is thought that one of the great white shark items now preserved in the latter museum represents the remaining part of the once mounted specimen (Larry Bell, pers. comm. 2008; Bigelow and Schroeder, 1948; Carolyn Kirdahy, pers. comm. 2008).

Museum of Comparative Zoology, Museum of Natural History, Harvard University

Harvard University, 26 Oxford Street, Cambridge, MA 02138
Specimens: 10.

The Museum of Comparative Zoology of Harvard University in Cambridge, Massachusetts, has in its catalogue materials belonging to ten specimens of great white shark. These materials are labeled with the following catalogue numbers: MCZ 89505, MCZ-775, MCZ-1227, MCZ-925, MCZ 36470, MCZ 39719, MCZ 153575, MCZ 153626, MCZ 153627, MCZ 164195.

Catalogue no. MCZ 89505 was preserved in liquid (alcohol). The great white shark was captured in June 1848, in the Gulf of Maine, off Provincetown, Cape Cod, Massachusetts. It measured 396.2 cm (13 ft) TL and its weight was estimated at about 680.4 kg (1500 lb). This item was acquired by the Harvard University Museum of Comparative Zoology in 1862 and is currently missing. The collector was Capt. N.E. Atwood, and the determiners were Karsten E. Hartel and Guido Dingerkus. This specimen was described by David Humphreys Storer as a new species, *Carcharias atwoodi*. Therefore, this specimen is the holotype of *Carcharias atwoodi*, synonym of *Carcharodon carcharias*. The jaws under MCZ-775 may belong to this specimen. This specimen was described in 1848, but only cataloged with a "holding" number in 1991. The specimen could not be found in 1991, but was cataloged under this number based on the original description, in the hope that the definite holotype would be found. Karsten E. Hartel researched the problem in 1986 and suggested that the specimen was actually MCZ-775, probably based on specimen locality and collector, but he found no clear link to the original description (Andrew Williston, pers. comm. 2008; Bigelow and Schroeder, 1948; Boston Society of Natural History, 1851; Ichthyology Collection Database of the Harvard University Museum of Comparative Zoology, 2008).

The items with catalogue no. MCZ-775 are a chondrocranium and a set of jaws. The chondrocranium and jaws were from a great white shark that was captured in the Gulf of Maine, off Cape Cod, Massachusetts. These items were acquired by the Harvard University Museum of Comparative Zoology in 1862. The collector was Capt. N.E. Atwood. Karsten E. Hartel and Dingerkus (1986, MS) state this chondrocranium and jaws might have been from the holotype of *Carcharias atwoodi* (catalogue no. MCZ 89505). It was cataloged as being only skull or jaws. These jaws have not been found and the reason they are missing is unknown (Andrew Williston, pers. comm. 2008; Ichthyology Collection Database of the Harvard University Museum of Comparative Zoology, 2008).

Catalogue no. MCZ-1227 was preserved in liquid (alcohol). The great white shark was captured in the Gulf of Maine, off Provincetown, Cape Cod, Massachusetts. This item was acquired by the Harvard University Museum of Comparative Zoology in the mid–1800s. The item is currently missing. This specimen was "Dried and prepared for mounting" after it was cataloged. Nothing more is known about this specimen (Andrew Williston, pers. comm. 2008; Ichthyology Collection Database of the Harvard University Museum of Comparative Zoology, 2008).

Catalogue no. MCZ-925 was preserved in liquid (alcohol). The great white shark was captured off Massachusetts and is currently missing. Nothing is known about its disappearance (Andrew Williston, pers. comm. 2008; Ichthyology Collection Database of the Harvard University Museum of Comparative Zoology, 2008).

The items with catalogue number MCZ 36470 are a head, pectoral fins, pelvic fins, tail (location: ID. Coffin #5), and dorsal fin (location: TANKtemp), which were preserved in liquid (alcohol). The head, pectoral fins, pelvic fins, tail and dorsal fin were from a male great white shark that was captured on November 15, 1945, off La Jolla, just north of Scripps Pier, California (32°52.0'N, 117°15.0'W). The depth of water was approximately 9.1 m and it was caught on a setline. The head length is 42 cm, DUJP

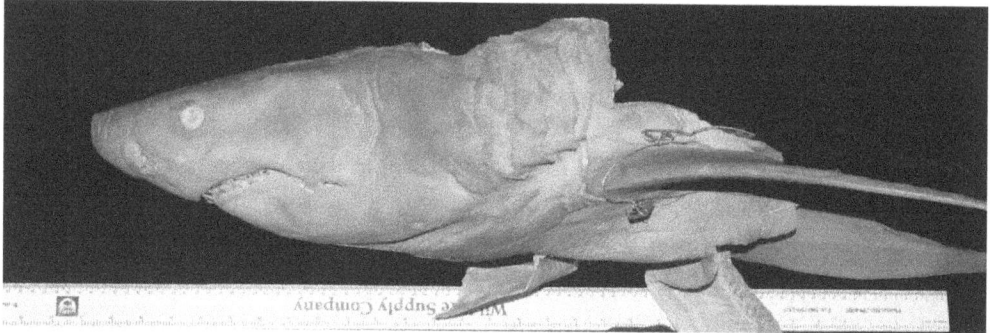

Head of a male great white shark captured on November 15, 1945, off La Jolla, California, and preserved at the Museum of Comparative Zoology of Harvard University in Cambridge, Massachusetts, with cat. no. MCZ 36470 (photograph courtesy Andrew Williston, Museum of Comparative Zoology, Harvard University).

is 28.4 cm and the teeth measure 1.292 cm UAE1 and 1.764 cm UAE2. The collector was G.L. Schillriff. Although catalogued, the tail is missing. The right side of the head was dissected to the skeleton. This specimen was acquired by the California Academy of Sciences from the Scripps Institution of Oceanography (where it was labeled with catalogue no. SIO 45-197) (Andrew Williston, pers. comm. 2008; Ichthyology Collection Database of the Harvard University Museum of Comparative Zoology, 2008; Ichthyology Collection Database of the Scripps Institution of Oceanography, 2008; H.J. Walker, Jr.).

Catalogue no. MCZ 39719 is a whole specimen, which was preserved in liquid (alcohol). The tail was separated from the body and stored together with the body (location: Coffin #3). The male great white shark was captured in 1948, in the Gulf of Maine, near Boston Lightship, Massachusetts. This specimen measured 151.7 cm TL, the DUJP is 21.0 cm and the teeth measure approximately 1.4 cm UAE1 and approximately 1.6 cm UAE2. The collector was John W. Lowes (Andrew Williston, pers. comm. 2008; Ichthyology Collection Database of the Harvard University Museum of Comparative Zoology, 2008).

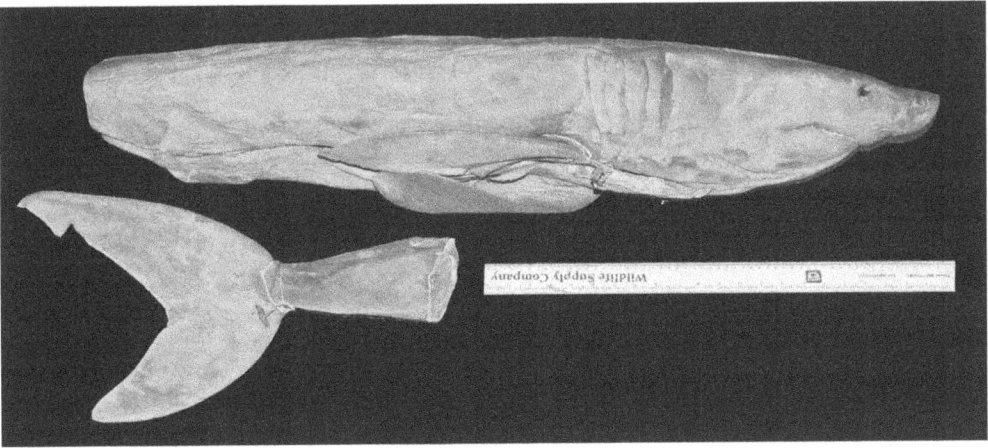

A 151.7 cm male great white shark that was captured in 1948, in the Gulf of Maine, near Boston Lightship, Massachusetts, and preserved at the Museum of Comparative Zoology of Harvard University in Cambridge, Massachusetts, with cat. no. MCZ 39719 (photograph courtesy Andrew Williston, Museum of Comparative Zoology, Harvard University).

Catalogue no. MCZ 153575 is a half lower jaw, which was preserved dry. The determiner was Guido Dingerkus. This dried specimen was cataloged in 1998, but it is certainly a much older specimen. Until the late 1990s skeletal material in the ichthyology collection was not consistently cataloged or was sometimes poorly labeled. Consequently, many unlabeled skeletons had to be given new catalog numbers. So it seems possible that this jaw is one of the three older "missing" MCZ *Carcharodon* specimens. However, this thought is speculation and there is currently no conclusive evidence supporting this theory (Andrew Williston, pers. comm. 2008; Ichthyology Collection Database of the Harvard University Museum of Comparative Zoology, 2008).

Catalogue no. MCZ 153626 is a set of jaws, which was preserved dry. The DUJP is 68.3 cm and the teeth measure 2.761 cm UAE1 and 3.021 cm UAE2. The determiner was William C. Schroeder. This item should be the one reported in Bigelow and Schroeder (1948) as being from a specimen of about 365.8 cm (12 ft) from an unknown locality. In Bigelow and Schroeder (1948) this item is cited without a catalogue number. Like the half jaw with catalogue no. MCZ 153575, this dried specimen was cataloged in 1998, but it is certainly a much older specimen. So it seems possible that these jaws are in fact one of the three older "missing" MCZ *Carcharodon* specimens. Again, this thought is speculation (Andrew Williston, pers. comm. 2008; Bigelow and Schroeder, 1948; Ichthyology Collection Database of the Harvard University Museum of Comparative Zoology, 2008).

Catalogue no. MCZ 153627 is a set of jaws. The DUJP is 61.0 cm and the teeth measure 2.411 cm UAE1 and 2.677 cm UAE2. The determiner was Karsten E. Hartel (Andrew Williston, pers. comm. 2008; Ichthyology Collection Database of the Harvard University Museum of Comparative Zoology, 2008).

Catalogue no. MCZ 164195 is a chondrocranium, which was preserved in

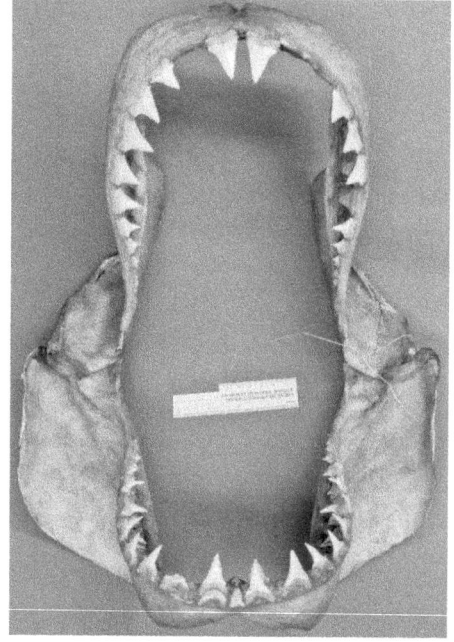

Set of jaws from a specimen of about 365.8 cm from an unknown locality, preserved at the Museum of Comparative Zoology of Harvard University in Cambridge, Massachusetts, with cat. no. MCZ 153626 (photograph courtesy Andrew Williston, Museum of Comparative Zoology, Harvard University).

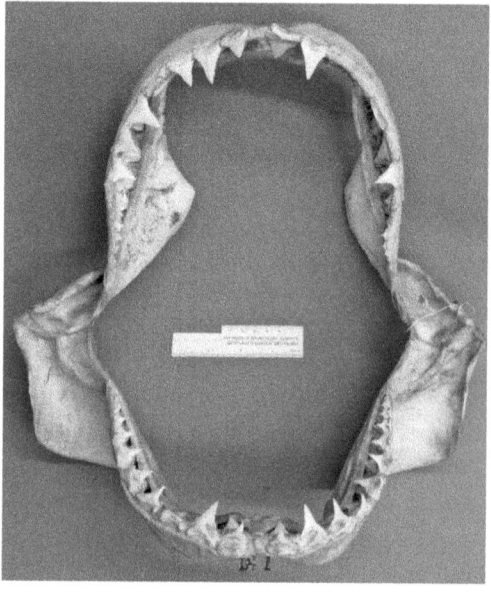

Set of jaws preserved at the Museum of Comparative Zoology of Harvard University in Cambridge, Massachusetts, with cat. no. MCZ 153627 (photograph courtesy Andrew Williston, Museum of Comparative Zoology, Harvard University).

liquid (alcohol). The chondrocranium was from a great white shark that was captured on December 7, 2004, 2.9 miles due east of Cape Charles Light on Smith Island, Virginia. This specimen measured 260 cm TL. It was collected by commercial fisherman inside the Virginia three mile limit. The collector was Capt. Paul Heric aboard the fishing vessel *Sundowner*, and the determiner was Shelton P. Applegate. The set of jaws and eyes of this specimen are preserved in the Virginia Institute of Marine Science in Gloucester Point, Virginia, and are labeled with catalogue no. 11419, but strangely these items have a different collection date of June 1, 2006 (Andrew Williston, pers. comm. 2008; Ichthyology Collection Database of the Harvard University Museum of Comparative Zoology, 2008; Ichthyology Collection Database of the Virginia Institute of Marine Science, 2008).

Berkshire Museum

39 South Street, Pittsfield, MA 01201
Specimens: 1.

The Berkshire Museum in Pittsfield, Massachusetts, has in its catalogue materials belonging to a single specimen of great white shark. This material is without a catalogue number.

A fiberglass mount of a great white shark head, with actual great white shark teeth in most of the mouth, on display at the Berkshire Museum in Pittsfield, Massachusetts (without cat. no.) (photograph courtesy Scott LaGreca, Berkshire Museum, Pittsfield, Massachusetts).

The item without catalogue number is a fiberglass mount of a great white shark head, with actual great white shark teeth in most of the mouth. The teeth measure 3.33 cm UAE1, 3.81 cm UAE2. The shark head is about 40–50 years old and is currently on display in the Berkshire Museum shop (Scott LaGreca, pers. comm. 2008).

NEW YORK

American Museum of Natural History

Central Park West at 79th Street, New York, NY 10024-5192
Specimens: 7.

The American Museum of Natural History in New York City, New York, has in its catalogue materials belonging to seven great white shark specimens. These materials are labeled with the following catalogue numbers: without catalogue number, 57269, 225701, 16493, 55200, 14625, 53095.

Catalogue no. 14773 is a set of jaws from a specimen about 182.9 cm (6 ft) long from Long Island, New York, which was reported in Bigelow and Schroeder (1948) as being preserved in the museum. On page 136 of Bigelow and Schroeder (1948), an illustration of the teeth of this specimen by zoological artist E.N. Fischer was also featured. Today this item is not listed in the museum's catalogue and could not be found (Barbara A. Brown, pers. comm. 2009).

Cast replica made from a 452.7 kg great white shark caught in 1935 off Manasquan, New Jersey, on exhibit to the public in the Milstein Hall of Ocean Life, at the American Museum of Natural History in New York City, New York (photograph by Gianluca Cugini, courtesy of the American Museum of Natural History, New York).

The item without catalogue number is a cast, which may have real teeth. This cast measures approximately 274.3–304.8 cm (9–10 ft) TL. According to Richard Ellis, this replica was made from a cast of a 452.7 kg (998 lb) specimen caught by Francis Hine Low in 1935 off Manasquan, New Jersey. In his 1963 book, *Fishing Is for Me*, Low wrote, "It was the first taken in this country on rod and reel, and the largest fish of any kind caught by an individual." Low was a trustee of the American Museum of Natural History. The item is currently on exhibit to the public in the Milstein Hall of Ocean Life (R. Ellis, pers. comm. 2009; Stephen C. Quinn, pers. comm. 2009).

Catalogue no. 57269 was preserved in liquid (alcohol). The great white shark was captured on July 6, 1985, 27 miles south east of Moriches Inlet, Suffolk, Long Island, New York. The collectors were M. Hagerott and M.N. Feinberg (Ichthyology Collection Database of the American Museum of Natural History, 2008).

The items with catalogue number 225701 are some of the fins, which were preserved in liquid (alcohol). The fins were from a great white shark that was captured on June 14, 1980, 10–15 miles south-southeast of center Moriches, Suffolk, Long Island, New York. The collector was Guido Dingerkus (Barbara A. Brown, pers. comm. 2009; Ichthyology Collection Database of the American Museum of Natural History, 2008).

Catalogue no. 16493 is a single tooth, which was preserved dry. The tooth was from a great white shark that was captured on May 30, 1945, off Jones beach in Nassau, Long Island, New York. The collector was F.H. Ketcham (Barbara A. Brown, pers. comm. 2009; Ichthyology Collection Database of the American Museum of Natural History, 2008).

Catalogue no. 55200 is an almost whole specimen (jaws missing), which was preserved in liquid. The great white shark was captured in September 1982, about 7 miles south of Montauk, Suffolk, Long Island, New York. The collectors were I. Klein and J. Casey (Barbara A. Brown, pers. comm. 2009; Ichthyology Collection Database of the American Museum of Natural History, 2008).

Catalogue no. 14625 was preserved in liquid. The great white shark was captured on May 24, 1939, off Ocean Beach in East Hampton, Suffolk, Long Island, New York. The collector was W.T. Helmuth (Ichthyology Collection Database of the American Museum of Natural History, 2008).

Catalogue no. 53095 is a set of jaws, which was preserved dry. The jaws were from a great white shark that was captured in June 1982, off Hobart, Tasmania, Australia. The collectors were fishermen and P. Lewis (Barbara A. Brown, pers. comm. 2009; Ichthyology Collection Database of the American Museum of Natural History, 2008). It seems likely that this great white shark is the one in the account that follows. On a morning in the winter of 1982, professional fishermen Colin Pettman discovered a dead female great white shark entangled in a graball net he had set on a rocky bottom in around 50 meters of water, approximately 7.5 kilometers north of Eaglehawk Neck, near Deep Glen Bay, Tasmania. The capture location is approximately 75 km south of

Hobart. In Black (2008) the capture is recorded as occurring around the middle of July, but the fisherman was unsure of the exact month, so it could have been June (Chris Black, pers. comm. 2009). The great white shark measured approximately 300 cm TL, and appeared to be in very poor physical condition, with a noticeably narrow girth. The stomach was full of sponges. Whether this is a symptom of starvation (it was a cold winter and fish were very scarce), or some other serious ailment, the ingestion of sponges is atypical of large predatory sharks. Black (2008) suggested that the shark inadvertently swallowed the sponges during its death throes. It is possible that the seafloor at the capture location supported sponge beds, and as the shark struggled in vain against the resistance of the net, it would have gaped repeatedly in an attempt to oxygenate its gills. The fisherman sold the whole shark carcass for processing, and it is quite likely that the jaws were salvaged at that point and could have found their way to New York via a collector. "P. Lewis," who is mentioned in the Ichthyology Collection Database of the American Museum of Natural History, may be Peter Lewis, a shark enthusiast in Tasmania at the time (Black, 2008; Chris Black, pers. comm. 2009).

NORTH CAROLINA

North Carolina Maritime Museum

315 Front Street, Beaufort, NC 28516-2125
Specimens: 1.

The North Carolina Maritime Museum in Beaufort, North Carolina, has in its catalogue materials belonging to one great white shark specimen. This material is without catalogue number.

The item without catalogue number is a model of the front half of a specimen that was molded from the actual shark. The male great white shark was captured on September 26, 1984, about 50 miles south of Beaufort Inlet, North Carolina. This specimen measured 472.4 cm (15.5 ft) TL and weighed 943.5 kg (2080 lb). The collectors were Lloyd Davidson, Jon Dodrill and Sylvester Karasinski aboard the fishing boat *Alligator*. The item is currently on exhibit to the public, hanging from a wall immediately to the left of the doorway entrance. Lloyd Davidson, Jon Dodrill and Sylvester Karasinski caught the shark on September 26, 1984. Davidson, captain of the 41' commercial fishing boat, *Alligator*, and his crew were fishing approximately 40 miles south of Cape Lookout in 28 fathoms of water. They set two separate one mile long 130 hook longlines baited predominantly with whole fresh striped mullet, beginning late in the day of the 25th and continuing after dark. The two lines as Jon Dodrill recalls were roughly a mile apart and set at the same approximate depth. This was the evening of September 25, 1984, about 10 days after a major hurricane had moved through the area. The temperature of the water was approximately 70° F. The men were longlining for sharks, and had already pulled in a haul that included several tiger sharks *Galeocerdo cuvier*. The great white shark did not get hooked on the line until the morning of September 26, 1984. It was still alive when the fishermen saw it. The great white had been feasting on sharks that had already been hooked, and it became tangled in the line. Later, they found out how they had captured this shark, as it had swallowed whole a 182.9 cm (6 ft) tiger shark that they had hooked. It had a hook in its belly, a hook in its mouth, and the tail was wrapped in the cable. The great white shark stomach also contained at least two very fresh severed heads of adult sandbar sharks *Carcharhinus plumbeus*. The cuts at the base of the head were straight knife cuts. Jon Dodrill believes that the white shark had picked up the discarded heads from sandbar sharks they captured earlier and discarded overboard from their first line. They moved to the second line where the white shark had wrapped its tail in the cable mainline, after it had swallowed the hooked 182.9 cm (6 ft) tiger shark whole. It had also hooked itself in the jaw on a second hook, a giant double-strength tuna hook that was on the adjacent gangion from where the tiger shark had been hooked. The fishermen wrapped a heavy line around the shark, then tied it off to the back of their boat. The weather was calm, the seas flat and clear, but weather conditions were deteriorating as a storm was moving in, so Davidson decided to return to port early. The great white would not fit on the *Alligator*, so they pulled as much of it on board as possible, and towed the rest. With their catch thus secured, the Alligator returned to port in Morehead City, where it arrived sometime after midnight on September 27, 1984. The shark was weighed on a truck scale at 943.5 kg (2080 lb) and examined on that day by Frank Schwartz of the University of North Carolina Institute of Marine Science in Morehead City, North Carolina. Dodrill believes he removed the heart. Dodrill and colleagues sent a vertebral sample for aging to Jack Casey at the National Marine Fisheries Service Laboratory in Narragansett, Rhode Island. The shark's carcass went on display at Ottis' Fish Market in Morehead City, where it remained for the weekend. It was estimated that over 3,500 people visited the Morehead City waterfront to see the large shark. At the end of the weekend, the shark went into a freezer to preserve it for future showings. A special exhibit was built to show the shark at the North Carolina State Fair in Raleigh. Along with the remains of the shark, drawings and informational handouts were prepared to educate people about sharks of the Carolina waters. By the time the fair ended, over 50,000 people had come by to take a look at the great white. Later, taxidermist Al Swanson of Virginia Beach, Virginia, mounted half of the shark (the back third of the shark was shipped to New York and sold as food at $2.00/pound). The Bill Walker family, who owned the trophy, donated it to the North Carolina Maritime Museum in Beaufort. The jaws were removed, and later the teeth were removed from the jaw. The North Carolina State Museum of Natural Sciences in Raleigh, North Carolina, holds various parts of this specimen (catalogue no. NCSM 27589 and NCSM 28360) (Jon Dodrill, pers. comm. 2009; Hairr, 2004; Wayne C. Starnes, pers. comm. 2008).

North Carolina State Museum of Natural Sciences
MSC# 1626, Raleigh, NC 27699-1626
Specimens: 2.

The North Carolina State Museum of Natural Sciences in Raleigh, North Carolina, has in its catalogue materials belonging to two specimens of great white shark. These materials are labeled with the following catalogue numbers: NCSM 27589, NCSM 28360, NCSM 28427.

Catalogue no. NCSM 27589 (previous catalogue number: UNC-IMS 16812) is a specimen divided into various parts, which were preserved in liquid (70 percent ethyl alcohol). From the same specimen is a heart, with catalogue no. NCSM 28360 (previous catalogue number: UNC-IMS 16798), which was preserved in liquid (70 percent ethyl alcohol). The male great white shark was captured on September 26, 1984, approximately 50 miles south of Beaufort Inlet, North Carolina. The claspers measured 72.0–80.0 cm and the heart 22.0 cm. This item was acquired by the North Carolina State Museum of Natural Sciences from the University of North Carolina Institute of Marine Science in 1996. It was stored for many years in steel tanks that rusted and thus imparted a reddish brown coloration to this specimen. The North Carolina Mar-

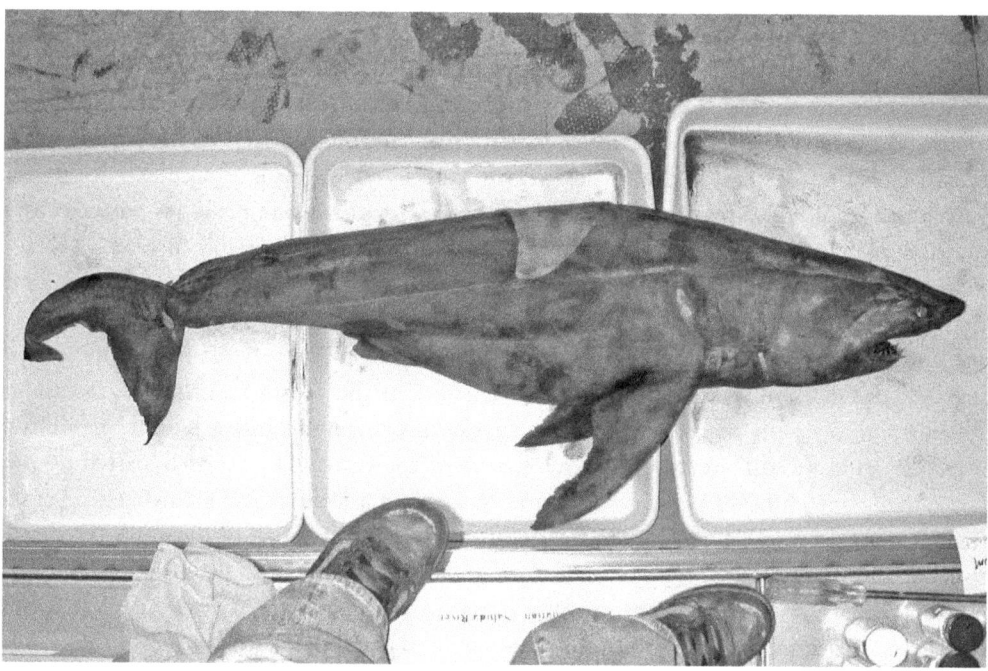

A 203.0 cm TL great white shark caught on April 18, 1974, off Shackleford Banks, North Carolina. It is preserved at the North Carolina State Museum of Natural Sciences in Raleigh, North Carolina, with cat. no. NCSM 28427 (photograph by Gabriela Hogue, courtesy North Carolina State Museum of Natural Sciences).

itime Museum in Beaufort, North Carolina, holds a model of the front half of this specimen, which was molded from the actual shark (without catalogue number). For a full description of this specimen's capture, see the section for the North Carolina Maritime Museum, Beaufort, North Carolina (Jon Dodrill, pers. comm. 2009; Hairr, 2004; Wayne C. Starnes, pers. comm. 2008).

Catalogue no. NCSM 28427 (previous catalogue number: UNC-IMS 9302) is a whole specimen, which was preserved in liquid (70 percent ethyl alcohol). The great white shark was captured on April 18, 1974, 0.5 miles off Shackleford Banks, about 5.9 air miles SE of center Beaufort, North Carolina (approx. 34.6582° N, 76.5913° W). This specimen measured 203.0 cm TL, 176.8 cm FOR and the teeth measure 2.35 cm UAE1. It was caught on a longline and was acquired by the North Carolina State Museum of Natural Sciences from the University of North Carolina Institute of Marine Science. The collectors were George H. Burgess, G.W. Link *et al.* This great white shark had eaten a dusky smooth-hound *Mustelus canis* (Mitchell, 1815), which had been hooked. Like the other specimens, it was stored for many years in steel tanks that rusted and thus imparted a reddish brown coloration to this specimen (Wayne C. Starnes, pers. comm. 2008).

PENNSYLVANIA

North Museum of Natural History & Science
400 College Ave., Lancaster, PA 17603-3393
Specimens: 1.

The North Museum of Natural History & Science in Lancaster, Pennsylvania, has in its catalogue materials belonging to a single specimen of great white shark. These materials are without a catalogue number.

The item without catalogue number is a set of jaws. The jaws were from a male great white shark that was captured off Albany, Western Australia. This specimen measured 518.2 cm (17 ft) TL and weighed 1814.4 kg (4,000 lb). The DUJP is 92.0 cm and the teeth measure 3.7 cm UAE1 and 3.9 cm UAE2. It was captured in gill nets set for smaller sharks. This item was acquired by the North Museum of Natural History & Science on April 9, 1987. The collector was Stuart Cramer and this specimen was prepared by Gordon Hubbell (Metro Zoo Miami) (Alison Eichelberger, pers. comm. 2008).

Academy of Natural Sciences
1900 Benjamin Franklin Parkway, Philadelphia, PA 19103-1101
Specimens: 4.

The Academy of Natural Sciences in Philadelphia, Pennsylvania, has in its catalogue materials belonging to four specimens of great white shark. These materials are

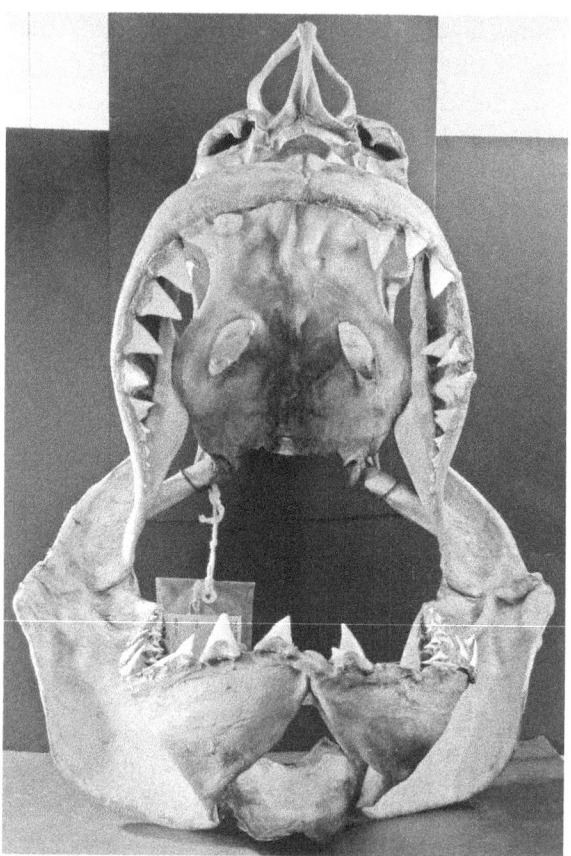

The chondrocranium, set of jaws and hyoid arch from a great white shark captured off Provincetown, Massachusetts, and preserved at the Academy of Natural Sciences in Philadelphia, Pennsylvania, with cat. no. ANSP 69984 (photograph courtesy Academy of Natural Sciences, Philadelphia).

labeled with the following catalogue numbers: ANSP 157328 , ANSP 69984, ANSP 50982, and ANSP 69986.

The items with catalogue no. ANSP 157328 are 8 upper and lower teeth, which were preserved dry. The teeth were from a great white shark that was captured off Englewood, Florida. This specimen measured 472.4 cm (15 ft 6 in). The teeth come from the A.R. Cahn Collection. The item is not currently on exhibit to the public (Ichthyology Collection Database of the Academy of Natural Sciences in Philadelphia, 2008). These teeth should be from the 470 cm immature female reported in Springer (1939). This specimen was taken by Fred Dalton and O.E. Holly of the Bass Biological Laboratory on one of the ordinary shark set-lines located about 8 miles off Englewood, Florida, on February 1, 1939. The capture was accidental in that the shark had wrapped the main chain of the set-line around its tail before it straightened the hook through its struggle. The animal was still alive when it was brought to the beach four days after capture. The stomach of this specimen contained two sandbar sharks *Carcharhinus plumbeus*, each between 182.9 cm (6 ft) and 213.3 cm (7 ft) long, and both showing evidence of having been torn from hooks on the set-line (Springer, 1939). Morphometric measurements taken from the white shark specimen were reported in Springer (1939).

The items with catalogue no. ANSP 69984 are a chondrocranium, a set of jaws, and a hyoid arch which were all preserved dry. The chondrocranium and jaws were from a great white shark that was captured off Provincetown, Massachusetts. The collectors were personnel from the Brooklyn Museum and the determiner was Henry W. Fowler. This item is not currently on exhibit to the public (Ichthyology Collection Database of the Academy of Natural Sciences in Philadelphia, 2008).

Catalogue no. ANSP 50982 is a set of jaws that was preserved dry. The jaws were from a great white shark that was captured in July 1920, off Beach Haven, New Jersey. The collector was H.S. Drinker, and the determiner was Henry W. Fowler. This item is not currently on exhibit to the public (Ichthyology Collection Database of the Academy of Natural Sciences in Philadelphia, 2008).

Catalogue no. ANSP 69986 is a set of jaws, which was preserved dry. The jaws were from a great white shark that was captured on December 15, 1939, off Arcadia Plantation, South Carolina. The collector was George Vanderbilt, and the determiner was Henry W. Fowler. This item is not currently on exhibit to the public (Ichthyology Collection Database of the Academy of Natural Sciences in Philadelphia, 2008).

SOUTH CAROLINA

National Ocean Service Marine Forensics Branch, Center for Coastal Environmental Health and Biomolecular Research

219 Ft. Johnson Road, Charleston, SC 29412-9110
Specimens: 4.

The National Ocean Service Marine Forensics Branch in Charleston, South Carolina, has in its catalogue materials belonging to four specimens of great white shark. These materials are labeled with the following catalogue numbers: 19-47-13, 20-60(22), 20-60-9, Ccar002b(106), Ccar003b(217s).

Catalogue no. 19-47-13 is a muscle, which was preserved frozen. The muscle was from a great white shark that was captured off Narragansett, Rhode Island. This specimen measured 444 cm FOR and weighed 1319.5 kg (2909 lb). This item was acquired by the National Ocean Service Marine Forensics Branch on May 11, 1991. This item is not currently on exhibit to the public (Julie Carter, pers. comm. 2008).

The items with catalogue no. 20-60(22) and 20-60-9 are a muscle, which was preserved frozen, and a skin, which was preserved in liquid (ethyl alcohol). The muscle and skin were from a great white shark that was captured on May 17, 1993, off Orange Grove, Florida. This item was acquired by the National Ocean Service Marine Forensics Branch on May 24, 1993. This item is not currently on exhibit to the public (Julie Carter, pers. comm. 2008).

Catalogue no. Ccar002b(106) is a muscle, which was preserved frozen. The muscle was from a female great white shark that was captured on February 22, 1994, off St. Petersburg, Florida. This specimen measured 391 cm TL. This item was acquired by the National Ocean Service Marine Forensics Branch on April 10, 1995, and is not currently on exhibit to the public (Julie Carter, pers. comm. 2008).

Catalogue no. Ccar003b(217s) is a steak, which was preserved frozen. The steak was from a great white shark that was captured on January 5, 1996 (Loran C, 45365,60580). This item was acquired by the National Ocean Service Marine Forensics Branch on January 8, 1996, and is not currently on exhibit to the public (Julie Carter, pers. comm. 2008).

TEXAS

Houston Museum of Natural Science
One Hermann Circle Drive, Houston, TX 77030
Specimens: 1.

The Houston Museum of Natural Science in Houston, Texas, has in its catalogue materials belonging to one specimen of great white shark. This material is labeled with the following catalogue numbers: 1986.781.

Catalogue no. 1986.781 is a fiberglass model of a great white shark that was captured off the coast of California. This specimen measured 396.2 cm (13 ft) TL and weighed 816.5 kg (1800 lb). It was caught with a harpoon. The molds were acquired by the Houston Museum of Natural Science from Sea World, San Diego, California, on September 8, 1986. The fiberglass model was made by Mr. Dwayne Hicks, an employee of the Houston Museum of Natural Science. The item is not currently on exhibit to the public (Dan Brooks, pers. comm. 2008).

VIRGINIA

Virginia Institute of Marine Science, College of William and Mary
P.O. Box 1346, Gloucester Point, VA 23062-1346
Specimens: 5

The Virginia Institute of Marine Science in Gloucester Point, Virginia, has in its catalogue materials belonging to five specimens of great white shark. These materials are labeled with the following catalogue numbers: VIMS 03562, VIMS 07358, VIMS 09602, VIMS 11343, VIMS 11419.

Catalogue no. VIMS 03562 is a set of jaws. The jaws were from a great white shark that was captured on May 29, 1957, off Sea Isle City, New Jersey. This specimen measured 229 cm (7.5 ft) TL. The item is not currently on exhibit to the public (Charles F. Cotton, pers. comm. 2008, 2009; Eric J. Hilton, pers. comm.; Ichthyology Collection Database of the Virginia Institute of Marine Science, 2008).

The items with catalogue number VIMS 07358 are five teeth. The teeth were from a great white shark that was captured on November 28, 1971, off Sandbridge, Virginia. It was caught midday, just beyond the surf line in a haul seine. This specimen measured 172.7 cm (5 ft 8 in). The collector was Don Lips. The determiner was Stewart Springer. The item is not currently on exhibit to the public (Eric J. Hilton, pers. comm.; Ichthyology Collection Database of the Virginia Institute of Marine Science, 2008).

The items with catalogue no. VIMS 09602 are teeth. The teeth were from a great white shark that was captured on February 2, 1999, off New Jersey. It was caught in a purse seine. This specimen measured approximately 120 cm TL. The collector was Jack A. Musick. This item is not currently on exhibit to the public (Charles F. Cotton, pers. comm. 2008, 2009; Ichthyology Collection Database of the Virginia Institute of Marine Science, 2008).

Catalogue no. VIMS 11343 is a whole specimen that was preserved in liquid (originally fixed in formalin, then stored in 70 percent ethyl alcohol). The male great white shark was captured on June 16, 2004, off Smith Island Shoals (GPS, 37.11.666 N, 75.71.666 W). It was caught on a Virginia Institute of Marine Science longline. This specimen measured 250 cm TOT, 218 cm TLn and 189 cm PCL. Fresh pinniped flesh and bones were found in the stomach by Charles F. Cotton. This item was acquired by the Virginia Institute of Marine Science on June 16, 2004. The collector was Jack A. Musick aboard the vessel *Bay Eagle*. The determiners were Jack A. Musick and J. Romine. This specimen is not "on public exhibit," being housed in a coffin. However, the Virginia Institute of Marine Science

A 250 cm TOT male great white shark that was captured on June 16, 2004, off Smith Island Shoals, Virginia, and preserved as a whole specimen at the Virginia Institute of Marine Science in Gloucester Point, Virginia, with cat. no. VIMS 11343 (photograph courtesy Virginia Institute of Marine Science, Gloucester Point).

staff routinely show it to tour groups that come to visit the Fish Collection of the Virginia Institute of Marine Science. Photos of this specimen were taken after capture, showing Charles F. Cotton performing necropsy, stomach contents, the mouth open and the weighing of the specimen (Charles F. Cotton, pers. comm. 2008, 2009; Ichthyology Collection Database of the Virginia Institute of Marine Science, 2008).

The items with catalogue no. VIMS 11419 are a set of jaws and eyes. The jaws and eyes were from a female great white shark that was captured on December 7, 2004, 3 miles east of Cape Charles Light, Smith Island, Virginia. This specimen was caught on a longline and measured 260 cm TL. The collector was Paul Heric aboard the fishing vessel *Sundowner*. The chondrocranium of this specimen is preserved in the Museum of Comparative Zoology of Harvard University in Cambridge, Massachusetts, and is labeled with catalogue no. MCZ 164195. The date of June 1, 2006, reported in the Collection Database of the Virginia Institute of Marine Science, is either the date of acquisition or perhaps the date that the loan was issued to the Museum of Comparative Zoology of Harvard University. This item is not currently on exhibit to the public (Charles F. Cotton, pers. comm. 2008, 2009; Ichthyology Collection Database of the Harvard University Museum of Comparative Zoology, 2008; Ichthyology Collection Database of the Virginia Institute of Marine Science, 2008; Andrew Williston, pers. comm. 2008).

Virginia Aquarium & Marine Science Center

717 General Booth Blvd., Virginia Beach, VA 23451
Specimens: 1.

The Virginia Aquarium & Marine Science Center in Virginia Beach, Virginia, has in its catalogue materials belonging to one specimen of great white shark. These materials are without a catalogue number.

The items without catalogue number are a set of jaws and fins that were preserved frozen. The jaws and fins were from a great white shark that was captured off Virginia. The DUJP is 31.3 cm and the teeth measure 1.5 cm UAE1 and 1.6 cm UAE2 (Natasha Seibel, pers. comm. 2008; Tim W. Scott, pers. comm. 2008).

WASHINGTON (STATE)

University of Washington Fish Collection

University of Washington, School of Aquatic & Fishery Sciences, 1122 NE Boat St, Seattle, WA 98105
Specimens: 1.

The University of Washington Fish Collection in Seattle, Washington, has in its catalogue materials belonging to one specimen of great white shark. These materials are labeled with the following catalogue number: UW 044031.

Catalogue no. UW 044031 is a set of jaws, which were preserved dry. The jaws were from a male great white shark that was captured on September 1, 1950, off Willapa Harbor, Washington. This specimen measured 440.0 cm TL and

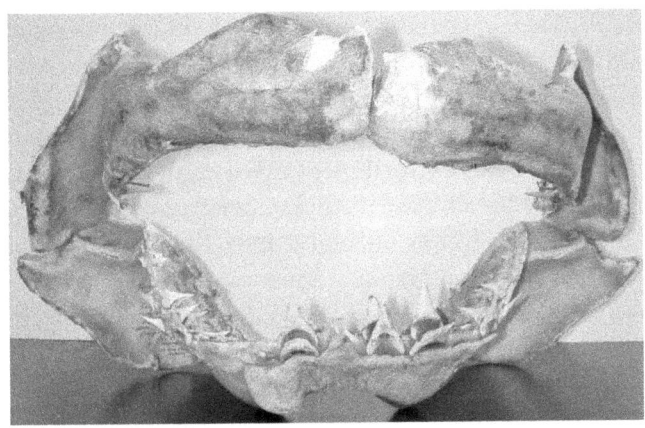

A set of jaws from a 440 cm male great white shark caught on September 1, 1950, off Willapa Harbor, Washington, and preserved in the University of Washington Fish Collection in Seattle, Washington, with cat. no. UW 044031 (photograph by Jeff Benca, courtesy University of Washington Fish Collection, Seattle, Washington).

was estimated to weigh 907.2 kg (2000 lb); the liver weighed 158.8 kg (350 lb). The DUJP is 91.5 cm and the teeth measure 33.7 cm UAE1 and 36.0 cm UAE2. It was caught in Mr. Raymond Nelson's gill net near buoy 22. The stomach contained four partly digested salmon *Oncorhynchus* sp., vertebral columns of North Pacific hake *Merluccius productus* (Ayres, 1855), rockfish *Sebastes* sp., the hides of two harbor seals *Phoca vitulina* (Linnaeus, 1758), and 150 crabs, primarily Dungeness crabs *Cancer magister* (Dana, 1852) with the remainder rock crabs *C. productus* Randall, 1839. The jaws from this shark were removed and sent to the School of Fisheries, University of Washington, where they were cleaned and dried by Arthur Donovan Welander for further study and exhibition. The functional row of teeth on the upper jaw had all been removed except for one near the center of the jaw (LeMier, 1951; Katherine Pearson Maslenikov, pers. comm. 2008).

WASHINGTON, D.C.

Smithsonian Institution National Museum of Natural History

10th Street and Constitution Ave. NW, Washington, D.C. 20560
Specimens: 9.

The Smithsonian Institution National Museum of Natural History in Washington, D.C., has in its catalogue materials belonging to nine specimens of great white shark. These materials are labeled with the following catalogue numbers: USNM 111185, USNM 170498, USNM 316465, USNM 196669, USNM 232642, USNM 232647, USNM 111168, USNM 110889, USNM 221663.

The items with catalogue no. USNM 111185 are teeth, which were preserved dry (location: NHB — North Attic). The teeth were from a great white shark that was captured in 1928, off Randall Cliff, Chesapeake Bay, Maryland. The collector was E. Reid. The item was lost as of August 1987 (Ichthyology Collection Database of the Smithsonian Institution National Museum of Natural History, 2008).

Catalogue no. USNM 170498 is a part of a specimen that was preserved in Tank 01. The great white shark was captured in the Mindoro Strait, Apo Light S. 65 Degrees W., 19.4 Miles, Philippines (GPS, 12°47'15" N; 120°41'00" E). The depth of water was 1068 m. The collectors were personnel aboard the vessel *Albatross* (Ichthyology Collection Database of the Smithsonian Institution National Museum of Natural History, 2008; Jeffrey T. Williams, pers. comm. 2009).

The items with catalogue no. USNM 316465 are teeth, which were preserved in liquid (alcohol). The teeth were from a great white shark that was captured on Aug 17, 1916, off Woods Hole, Massachusetts (Ichthyology Collection Database of the Smithsonian Institution National Museum of Natural History, 2008).

Catalogue no. USNM 196669 (location: Large Tank 10) is a whole specimen, which was preserved in liquid (originally frozen, then fixed in formalin, then stored in 75 percent ethyl alcohol). The great white shark, probably a female, was captured in the summer of 1961, 15 miles east of New Jersey (GPS, 39° 41' N; 73° 48' W). The depth of capture was between 0 and 18 m. This specimen measured approximately 243.8 cm (8 ft) TL. The collector and determiner was Victor G. Springer. He obtained this specimen on a NOAA longlining cruise, had it frozen and shipped to the Smithsonian Institution National Museum of Natural History (Ichthyology Collection Database of the Smithsonian Institution National Museum of Natural History, 2008; Victor G. Springer, pers. comm. 2009; Jeffrey T. Williams, pers. comm. 2009).

The items with catalogue no. USNM 232642 (previous catalogue number: USNM 27374) are a set of jaws and a tooth, which were preserved dry (location: NHB — North Attic). The jaws and tooth were from a great white shark that was captured off Soquel, California. The collector was D. Jordan. Four teeth were given (as USNM 27374) to the Natural History Museum of Los Angeles County on April 2, 1973 (Ichthyology Collection Database of the Smithsonian Institution National Museum of Natural History, 2008).

The items with catalogue no. USNM 232647 are vertebral discs, which were preserved dry (location: NHB — North Attic). The vertebral discs were from a great white shark that was captured off Sarasota, Florida (Ichthyology Collection Database of the Smithsonian Institution National Museum of Natural History, 2008).

Catalogue no. USNM 111168 is a tooth, which was preserved dry (location: NHB — North Attic). The tooth was from a great white shark that was captured on September 28, 1936, in Willapa Bay near North River, Washington. The collector was E. Pedersen (Ichthyology Collection Database of the Smithsonian Institution National Museum of Natural History, 2008).

Catalogue no. USNM 110889 (previous catalogue number: 11845) is a set of jaws, which was preserved dry (location: NHB — North Attic). The jaws were from a great white shark that was captured off Woods Hole, Massachusetts. This specimen measured 259.1 cm (8.5 ft) TL. The collector was S. Baird (Bigelow and Schroeder, 1948; Ichthyology Collection Database of the Smithsonian Institution National Museum of Natural History, 2008). This specimen is cited in Bigelow and Schroeder (1948) with the wrong catalogue number, 10899. The 10000 series of catalog numbers were assigned in the mid–1850s, and the penultimate number was clearly a typographical error (Jeffrey T. Williams, pers. comm. 2008). On page 135 of Bigelow and Schroeder (1948), an illustration of the teeth of this specimen by zoological artist E.N. Fischer was also featured.

Catalogue no. USNM 221663 are fibrous muscles from the tail, which were preserved dry. The muscles were from a great white shark that was captured off Eastport, Maine (Ichthyology Collection Database of the Smithsonian Institution National Museum of Natural History, 2008).

4

Great White Sharks in Aquariums (A Chronological Report)

Great White Sharks Are Not Suited for Captivity

The great white shark is a pelagic animal with a high metabolism, and it is not suited for captivity in aquariums. Currently, there is a single great white shark on exhibit at an aquarium. As of 2008, there had been 41 unsuccessful attempts to keep white sharks at public aquariums, dating back to the 1950s. In most cases these sharks did not feed, and were released after a few days or died. The reason for this lack of success was almost always related to the stresses associated with capture, transport and acclimatization to the facility (Hewitt, 1984). The longevity record for a white shark on exhibit is 198 days, established at the Monterey Bay Aquarium, in Monterey Bay, California, from September 2004 to March 2005. The Monterey Bay Aquarium is the only institution in the world to exhibit a white shark for more than 16 days (Monterey Bay Aquarium, 2008b). Dehart (2004) recognized the great white shark to be among the species that have specialized exhibit requirements, like very large exhibit dimensions in the horizontal plane and compatibility problems with other sharks. He concluded that communication with experienced institutions is strongly recommended before attempting to maintain this species. Smith *et al.* (2004), considering the limited success of transporting great white sharks, suggested that transport of individuals of this species should only be attempted by very experienced personnel.

The great white shark is a pelagic animal with a high metabolism and is not suited for captivity in aquariums. In this photo is a 147.3 cm male white shark that survived for 11 days at the Monterey Bay Aquarium, Monterey, California, in September 1984 (photograph courtesy Monterey Bay Aquarium).

Complete documentation available on all the cases in which great white sharks have been kept in captivity in U.S. aquariums is presented in this chapter. Since it is interesting to see how the keeping of white sharks has changed with time, occurrences are organized chronologically rather than by institution.

March 1955: Marineland of the Pacific, Palos Verdes Peninsula, California

In March 1955, a 152.4 cm (5 ft) newborn white shark was trucked from the Scripps Institution at La Jolla to Marineland of the Pacific, located at the tip of the Palos Verdes Peninsula in Los Angeles County, California. The shark survived only a few hours (Ellis and McCosker, 1991).

March 1961: Hawaii Marineland, Honolulu, Hawaii

On March 8, 1961, a 725.7 kg (1,600 lb), 406.4 cm (13 ft 4 in) long great white shark was caught in 27.4 m of water one mile outside Honolulu Harbor. Fred Inouye, skipper of the *Holokahana I*, and vice president of Hawaii Marineland, made the capture. The shark was brought to Hawaii Marineland in Kewalo Basin about noon of the same day. It was revived by two divers (Marineland employees Francis Warren and Gordon Kenolio), who held its pectoral fins and walked it in circles for about an hour and half to familiarize it with the circular tank. A photo showing the two Marineland employees walking the shark was published in the local newspaper *Honolulu Star Bulletin*. "As soon as it was revived, the men got out fast," said Stan Omiya, Marineland director. The shark apparently mistook portholes in the tank for openings and scraped its nose against ridges in the windows. On March 9 it appeared dead, and divers once more began walking it to revive it. Once revived, it made two laps around the tank and settled to the bottom again. Marine biologists were summoned to examine the shark. The specimen was displayed in the tank for 24 hours before it finally died (Anonymous, 1961a, 1961b; Taylor, 1985).

December 1962: Marineland of Florida, St. Augustine, Florida

In December 1962, a 2.4-m (8-ft) long male great white shark was captured off St. Augustine, Florida and brought to Marineland of Florida. According to the press release issued at that time, the shark survived 35 to 36 hours in captivity. After showing some signs of recuperating from its capture, the shark sank to the bottom of the tank and died (Ellis and McCosker, 1991).

May 1968: Sea World San Diego, San Diego, California

In 1968, the management at Sea World in San Diego, California, decided to try to keep blue sharks *Prionace glauca* and shortfin mako sharks *Isurus oxyrinchus* in captivity. As the curator of fishes, David C. Powell designed a relatively inexpensive, behind-the-scenes tank that he hoped would meet the needs of a pelagic creature like the blue shark. He designed a 65,000 gallon circular tank of 17.5 m diameter, and 1.0 m–2.1 m deep, with a sloping bottom that became shallower towards the outer perimeter. While the tank was under construction, they experimented with capture and transport methods. A fifty-pound burlap sack of ground mackerel was towed slowly behind the boat for half a mile or so, leaving a trail of odor that would smell like lunch to a blue shark. For transport, they had a 2.1-m-long-by-0.6-m-wide plastic-lined tank. Powell made a flattened plastic mouthpiece that fit inside the mouth of the shark. The mouthpiece had five holes on each side that theoretically lined up with the five gill slits of the shark. Through this device, water supersaturated with pure oxygen was pumped by a small submersible pump sitting in the narrow transport tank. For practice, they would capture the blue sharks and hold them in the long, narrow tank on board, nicknamed the "shark coffin," for a couple of hours and then release them. The behind-the-scenes holding tank at Sea World was finally ready and the water system turned on. "Gator" Bill Ervin and Powell went out a couple of miles off Mission Bay

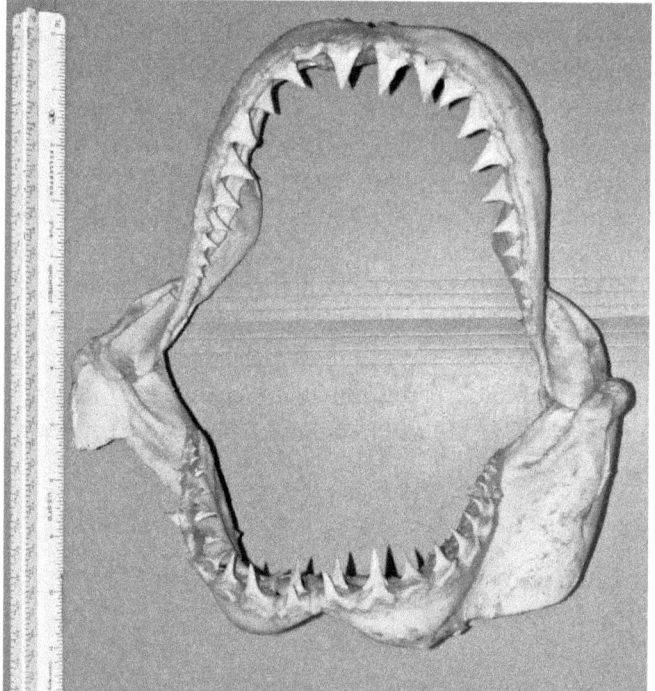

The set of jaws from a 180 cm white shark that survived for a little more than seven days at Sea World in San Diego, California, in May 1968 (photograph courtesy David C. Powell).

(where Sea World is located), laid their chum line of mackerel juice, and waited. Pretty soon a 1.8 m long, 54 kg shark showed up off the stern. Using a heavy nylon hand line, Powell tossed out a baited, barbless hook attached to a short wire leader. The shark quickly took the bait, and Powell instantly realized, as it nearly pulled him overboard, that this was no blue shark but most likely a much stronger mako. After finally getting it alongside the boat, Bill and Powell had a tough time lifting the heavy and uncooperative shark into the boat. Once it was lying upside down in the transport box, with the mouthpiece in place and the oxygen pump running, they took off for Sea World. Powell noticed that this particular shark repeatedly bit down on the mouthpiece, something he had not seen with the sharks during practice. Hoisting the "shark coffin" off the boat and driving it the hundred yards to the waiting shark tank, they released the shark and it swam off vigorously. The transport method seemed to have worked fine. The shark looked good as it cruised around the 15.2 m diameter tank. They spent the next eight days trying to get the shark to take food, but it refused everything they offered. Only once, when they poured a bucket of mackerel blood into the water directly ahead of it, did it show any response, but it still would not take the mackerel they dangled in front of it. On the eighth day it died, and only then did Powell realize that what they had caught was not a mako, but a young great white shark. In the excitement of catching and boating the shark, neither of them took the time to study its identifying features. Today, Powell still has the set of jaws of this specimen (Powell, 2001; David C. Powell, pers. comm. 2008, 2009).

October 1976: Steinhart Aquarium, California Academy of Sciences, San Francisco, California

On October 9, 1976, commercial halibut gillnetters Pete Halley and his brother caught a 213.4 cm (seven ft) great white shark in their net set over a sandy bottom in Bodega Bay, just outside the entrance of Tomales Bay. Through the marine radiotelephone operator, they called the Steinhart Aquarium, California Academy of Sciences, San Francisco. Steinhart Aquarium curator David C. Powell suggested that the two fishermen attach a line on the shark and tow it slowly back into Tomales Bay where Powell and Miller would meet them at the dock in Marshall, California. The Halley brothers managed to get a line on the shark and the shark swam behind the boat. Later, the line became wrapped in the propeller and they had to stop the boat and get the line free from the propeller. During this time, the shark was not able to swim and when they arrived at the dock in Marshall, it sank to the bottom lying on its side, showing no sign of life. Powell and aquarist Ed Miller arrived in Marshall with the collecting truck equipped with a 1.8 × 0.9 m fiberglass holding tank with an oxygen cylinder, regulator, oxygen pump and shark stretcher. However, when Powell and Miller put the shark into the fiberglass holding tank with highly oxygenated water, it started to move and breathe regularly. When they arrived at the Steinhart Aquarium, the shark was put in the Roundabout, a 100,000 gallon tank opened in May 1977, where pumps keep seawater constantly circulating. The Roundabout is toroidal or donut-shaped, with an 2.4-m wide, 3 m tall raceway that has a 20.1 m diameter. It was modeled after the donut-shaped exhibit at Shima Marineland, Japan, where viewers were located within the "donut hole" surrounded by a circle of schooling fishes, sharks, and rays. The shark started slowly swimming, but after a few strokes of the tail it stopped and sank to the bottom. Steinhart Aquarium director John McCosker picked the shark up to get it swimming again. The shark started swimming by itself, but stopped shortly and sank to the bottom. McCosker and Powell took turns picking the shark up to get it swimming again, but in the morning it died.

There is no doubt that this great white shark is the same specimen reported as having been captured on October 9, 1976, just outside the entrance to Tomales Bay, California, collected by Pete Halley and of which a gill arch, vertebrae, and other parts are preserved at the California Academy of Sciences with catalogue no. CAS 37917 (see page 128, "California Academy of Sciences") (Ichthyology Collection Database of the California Academy of Sciences, 2008; McCosker, 1999; Powell, 2001; David C. Powell, pers. comm. 2008, 2009).

1976–1977: Sea World San Diego, San Diego, California

In 1976 or 1977, curator John Rupp was directly involved with the attempted rehabilitation of a two meter great white shark at Sea World San Diego, California, prior to the construction of the Shark Encounter exhibit (see page 172, "1978–1982: Sea World San Diego"). This white shark was taken as bycatch from a seiner off the San Diego coast. John Rupp and Mike Shaw picked the shark up from the fishing boat. It was originally called in to Sea World by the captain, stating it was a shortfin mako shark *Isurus oxyrinchus*. Sea World curators had previously worked with mako sharks without success, so they were not terribly excited until they arrived at the boat and peered into the fish hold and saw a two-meter-long white shark cruising around in circles, very comfortably, in a fish hold that measured not greater than six ft square. The shark was transported to Sea World and placed into a 55,000 gallon rectangular tank, with a depth of approximately 4.3 m. This specimen survived for five or six days and never showed interest in food. The shark was thought to have died from a gaff wound under the pectoral fin (John Rupp, pers. comm. 2009).

1978–1982: Sea World San Diego, San Diego, California

Between 1978 and approximately 1982, six white sharks survived for various lengths of time at Sea World San Diego, San Diego, California. Ray Keyes, Mike Shaw and colleagues worked intently to display a live great white shark at Sea World San Diego. They tried to capture their own, but had to rely upon commercial gill net fisherman for the specimens that they worked with. The great white sharks were mostly caught off Oxnard, 50 miles north of Los Angeles, California. The Oxnard fish did not do well in that they were caught by commercial fishermen. The sharks did not receive the care that they needed, but were alive in their large Shark Encounter tank for several days. The Shark Encounter tank was a facility that the oceanarium staff designed and built for the display of large tropical sharks. This 40,000 gallon tank was designed with the help of David C. Powell, Jerry Klay and Ray Keyes. It opened to the public on June 19, 1978, and it was a leap forward at the time. The first white shark in the new aquarium arrived on July 4, 1978. The second one arrived when Ron and Valerie Taylor were in San Diego to work with the Sea World staff filming great white sharks. This second shark, which was approximately 167.6 cm (5.5 ft) long, was a male named "Ron." One specimen that was caught off San Diego by a commercial fisherman was in the best condition. According to Ray Keyes, it survived for some days but did not eat. On August 3, 1981, a 1.60–1.68 m long, 36 kg great white shark was kept in captivity for 16 days in the Shark Encounter tank. The white shark was injured as a result of its capture and never fed during its captivity. The great white shark died on August 19, 1981. At that point in time, the dead specimens were necropsied for further study. Ray Keyes did not believe that any of the specimens were preserved. The small white sharks gave no sign of behavioral distress or avoidance of the other sharks. Keyes, Shaw and colleagues had to be very careful about new shark introductions in that the bull sharks *Carcharhinus leucas* would harass and eat the new sharks. However, this was never the case with the small great white sharks. As Ray Keyes and Mike Shaw recall, they worked with at least six live small great white sharks at Sea World San Diego. They were all between 121.9 cm (4 ft 6 in) and 182.9 cm (6 ft) total length. Keyes and Shaw do not remember their sexes. They all survived transit from the capture location to Sea World. All six of these sharks were on display and were able to swim in the Shark Encounter tank, although they were damaged from being gill netted. From his experience, Ray Keyes came to believe that they were quite hardy (much more so than their close relative the shortfin mako *Isurus oxyrinchus*) in that they could survive the stress of rough capture followed by transit. None of these six great white sharks ever took food, although Keyes and colleagues tried to feed them. In order to meet the requirements of the bull sharks *Carcharhinus leucas* and the lemon sharks *Negaprion brevirostris*, the water temperature had been heated to approximately 75°F. The water was probably too hot for the juvenile great white sharks, and the presence of the adult bull sharks probably contributed to their distress. The white sharks appeared to be stressed from the constant swimming in the tank. There were times that their swim patterns were interrupted by their tank mates. The white sharks, despite their small size, appeared to dominate the display. All other sharks kept their distance. Keyes did not have specific dates for capture and death, but there should have been records at Sea World San Diego. However, these records are either hard to find or lost by this time. Suzanne M. Gendron recalls that the sharks did turn toward the wall and that was why they were constantly walked, day and night, from the surface of the aquarium with poles in hand to keep them from hitting the walls. She remembers that one shark was placed in the transport box and force fed. Four or five of the specimens died at Sea World and were examined. Tissue and blood samples were taken by veterinarian Lanny Cornell. At least one, and maybe two white sharks were released offshore after 3 days when their condition began to deteriorate (Ellis and McCosker, 1991; Suzanne M. Gendron, pers. comm. 2009; Hewitt, 1984; Ray Keyes, pers. comm. 2009; Mike Shaw, pers. comm. 2009).

August 1980: Steinhart Aquarium, California Academy of Sciences, San Francisco, California

In 1980, a female great white shark was kept in captivity for 3.5 days at the Steinhart Aquarium before being released. On August 12, 1980, Al Wilson, a fisherman who had once been employed as a specimen collector for the Bodega Bay Marine Laboratory, caught a 2.3 m (7.5 ft) long, 136.1 kg (300 lb) great white shark in his flounder net off Bodega Bay. Wilson slipped a rope around the shark's tail and towed it to shore. Wilson called the Steinhart Aquarium, which was offering a $US 1,000 reward to anyone who could deliver a living great white shark. The Steinhart Aquarium had a Steinhart White Shark Acquisition Team, which immediately arrived at Bodega Bay and brought the shark to the Steinhart Aquarium in a truck where aquarist Ed Miller massaged the shark and attended to its oxygen supply and water pumps. Here, the shark was kept in the Roundabout, a 100,000 gallon tank opened in 1977, where pumps keep seawater constantly circulating. The female great white shark was nicknamed "Sandy." The shark was light-sensitive so the light level was reduced and filters were introduced. Photographer Al Giddings dove in the tank and filmed the shark. The films were subsequently used to study the white shark swimming. While swimming, the shark started to hit against the four abutments of the tank, and, as the shark's health improved and its swimming speed increased, it hit the abutments with greater force. Chief designer Kevin O'Farrell solved the problem by installing a Plexiglas shield that smoothed out the inner surface of the tank. The aquarium staff tried to feed the shark with squid and many species of fish without success. By the fourth day of captivity, the shark would occasionally swim erratically through a five degree arc of the tank and collide with the outer wall at about the same point each time. Electrical engineer Norm Buell found that there was a minute electrical anomaly between two of the windows, a differential of 0.125 millivolt. To correct the electrical problem, they would have to drain the entire tank. Consequently, aquarium director John E. McCosker decided to release the shark. McCosker and his collaborators, including Al Giddings, took the shark to the Farallon Islands with the *Flying Fish*, a 16.8-m boat, and released it. During its captivity, the shark was seen by 40,000 visitors (Ellis and McCosker, 1991).

August 1981: Marineland of the Pacific, Palos Verdes Peninsula, California

In August 1981, Donald D. Zumwalt, Curator of Fishes at Marineland of the Pacific, received a marine telephone call from the sport fishing boat *Rebel* saying that they had taken a 58-inch-long great white shark. They were 10 miles off the Marineland coast and were coming into Marineland pier. In order to shorten transport time for the shark, Zumwalt immediately prepared their fiberglass 7.3 m long Boston Whaler with a transport box and launched it from the pier by davit. Zumwalt and his crew met the *Rebel* about 7 miles out and transferred the shark to their box. A crew was waiting on the pier with a truck and manning the hoist to lift the transport box onto the truck. It took only minutes to reach the base of the 540,000 gallon fish tank where another hoist crew were waiting to lift the shark and box directly into the tank. Divers were ready in the water to release the shark into the tank. The shark swam with ease and did not run into the walls or objects on the bottom. Feeding small fish to the shark was attempted, but with no results. After several days, it began to weaken, with a slower swimming rate and increased respiration. On the third day, it began running into the walls and spending time lying on the bottom. On the morning of the fourth day, he was found dead on the bottom of the tank. A necropsy was performed and the shark was in excellent condition except for gaff wounds in the gills and body (Donald D. Zumwalt, pers. comm. 2009).

Early '80s — August 1983: Steinhart Aquarium, California Academy of Sciences, San Francisco, California

After Sandy, the Steinhart Aquarium staff handled four other small white sharks in the early 1980s, the most recent being a 130 cm SL male specimen in August 1983 (John C. Hewitt, pers. comm. 2009; Hewitt, 1984). This great white shark is the same specimen reported as being captured on August 2, 1983, two miles due west of Ventura Marina entrance, California, collected by commercial fisherman Ben Henke and John C. Hewitt and preserved at the California Academy of Sciences with catalogue no. CAS 53045 (see page 128, "California Academy of Sciences") (John C. Hewitt, pers. comm. 2009; Hewitt, 1984; Ichthyology Collection Database of the California Academy of Sciences, 2008).

For several years, John E. McCosker and John C. Hewitt sought young white sharks for research purposes and hopefully for display at the Steinhart Aquarium. Their research involved studying the stress-induced changes in muscle and blood physiology of white sharks and other obligate ram ventilating sharks, with regards to their capture and transport. The majority of the animals were collected off Ventura California by local gill net fishermen. The main white shark nursery grounds for California are located inshore from Ventura Canyon off Southern California. Other smaller second and third year sharks were taken off central California in summer months near Bodega Bay, and Half Moon Bay. These white sharks were generally not young of the year, and were probably third year fish attaining 136.1 kg (300 pounds) and nearing 2.1 m (7 ft) in total length. The main goal of the research was to figure out a way to successfully capture and move great white sharks from the capture location to aquarium displays without killing them. This is not an easy task, unless there is a huge transport rig available in which they can swim somewhat normally. The Steinhart staff tried to keep all the sharks alive for exhibit, but some were very near death when they were received. Others were caught by fishermen, and the staff had to receive and transport the sharks near the capture location. John McCosker and John Hewitt posted "wanted posters" and spread the word among fishermen that if they should catch a white shark, the Steinhart staff wanted it. This strategy worked well, especially in Central California. McCosker and Hewitt learned a lot, and some of the sharks were exhibited in the Roundabout at Steinhart, but few did well in the long term. Cumulative stress of capture and poor transport practices were usually the cause of death. McCosker and Hewitt felt it would be preferable to actually catch one themselves. They worked about three or four years with commercial fisherman Ben Henke, who caught sharks regularly in his halibut gill net and shark nets. They would fish with him as deckhands, and if he caught one that was alive, it would be transported to the Steinhart Aquarium. A couple of times, Henke caught one in December when they were not prepared to work with it. The shark was either released or sold commercially for the meat. Of the four white sharks, some were caught by Ben Henke, and the others were caught by McCosker and Hewitt farther north during an El Niño year, which appeared to shift the young white sharks of the year to areas north of where they were normally found off Ventura and Santa Barbara. One of the sharks McCosker and Hewitt worked with was filmed extensively by underwater cinematographer Al Giddings, who at the time was making a television documentary called *Ocean Quest*.

John Hewitt still has a fiberglass model in his office of a young white shark that he caught with Ben Henke. It was dead on arrival at the boat. Hewitt took the jaws, which were included in this mount that was done by Lyons and O'Haver, Inc., of San Diego, California. Hewitt also has the jaw set from one of the 136.1 kg (300 lb) specimens that died, which is used for educational purposes (John C. Hewitt, pers. comm. 2009).

July 1984: Steinhart Aquarium, California Academy of Sciences, San Francisco, California

In July 1984 a great white shark survived for three days at the Steinhart Aquarium, California Academy of Sciences, San Francisco, California. In June and July of

1984, the Steinhart Aquarium returned to Ventura California in an attempt to capture another small white shark. After sixteen days of fishing in that area, a 1.5 m female specimen was obtained from a gill net fisherman. The animal had been captured approximately two miles off the beach in a monofilament halibut gill net, and was subsequently towed into Ventura harbor. Upon its arrival at the dock, it was tied by the tail and hung in the water (head down) for approximately four hours until staff biologists arrived on the scene. The animal was very stressed, and exhibited many open lesions and abrasions from the capture net and transport. It was exhibiting respiratory movements and swimming motion when stimulated, but was relatively weak and only moderately responsive. The animal was loaded into the transport vehicle, and fitted into a specially designed harness that provided a unidirectional current of superoxygenated water over the gills. The animal was subsequently transported back to the Steinhart Aquarium. Due to the mechanical breakdown of the transport vehicle and the necessity for frequent stops to check on the condition of the fish (only one biologist was present), the total elapsed time from obtaining the shark to arrival at the aquarium was 16 hours. However, the condition of the animal was much improved upon arrival at the aquarium and its condition improved continually during the first 36 hours of confinement, but the following 40 hours produced a gradual decline in vitality, and the white shark eventually died (Hewitt, 1984).

This great white shark is the same specimen reported as being captured on July 7, 1984, two miles west of the entrance to Ventura boat harbor, California collected by John C. Hewitt and preserved at the California Academy of Sciences with catalogue no. CAS 55435 (see page 128, "California Academy of Sciences") (Ichthyology Collection Database of the California Academy of Sciences, 2008).

September 1984: Monterey Bay Aquarium, Monterey, California

In September 1984, a white shark survived for 11 days at the Monterey Bay Aquarium, Monterey, California. On September 10, 1984, a 147.3-cm (4-ft 10-inch) long, 24.9 kg (55 lb) male white shark caught in a gill net by fisherman Joe Papetti off Bodega Bay, was placed in the Monterey Bay Habitats exhibit a little more than a month before the aquarium grand opening. Curator David C. Powell purchased the shark from the Bodega Bay fisherman and released it into the 345,000-gallon tank. The white shark had problems navigating the tank during the first two days, especially at night. Aquarists were stationed around the top of the tank to fend it off if it looked like it was going to hit the wall. After two days in captivity, it navigated the 27.4 m long exhibit well. At the time, the staff was extremely busy getting the whole aquarium ready for the opening and still had to perform regular maintenance on the tank. Needless to say, the volunteer divers and the staff were hesitant about diving in the tank. They had to be shown

In September 1984, this 147.3 cm male white shark survived for 11 days at the Monterey Bay Aquarium, Monterey, California (photograph courtesy Monterey Bay Aquarium).

that it really was safe, because only a few of them had dived with sharks before. Collector Bob Kiwala was the first one to dive with the white shark, and then cinematographer Mark Shelley of Sea Studios went in with his video camera. Powell went in to siphon the debris from the bottom of the tank. They tried everything they could think of to stimulate the shark to feed. At one point they suspended a live 61 cm long lingcod *Ophiodon elongatus* in front of its path. It turned to avoid it and swam on by. In the exhibit were a wide variety of fish including salmon, striped bass *Morone saxatilis*, several species of rockfish *Sebastes* sp., flatfishes and others. The shark ignored everything. As soon as the white was released into the tank, the four sevengill sharks that were already in the same tank before its arrival no longer cruised the length of the tank but went down to the deep end and stayed there until the white shark was gone. They were all larger than the white shark, yet the white shark had a dominating presence that the sevengills recognized. Powell decided to release the shark before it died. Unfortunately, it died on the day they were planning to release it, on September 20. The shark was in captivity for 11 days. After it died, it was stored in the Monterey Bay Aquarium freezer for some time and then turned over to Gregor M. Cailliet at Moss Landing Marine Laboratories in Moss Landing, California. At Moss Landing Marine Laboratories it was catalogued with catalogue no. B-35-0305, perhaps on January 15, 1986. Currently, this item is missing. The Moss Landing Marine Laboratories suffered serious damage from the 1989 Loma Prieta earthquake and the laboratories were torn down and reconstructed. All the ichthyology specimens were moved to other facilities and the California Academy of Sciences in San Francisco, California, was the only one large enough to take such a specimen. Therefore, it is presumed that this specimen or parts of it are among those currently preserved at the California Academy of Sciences (see page 128, "California Academy of Sciences") (Gregor M. Cailliet, pers. comm. 2009; Ellis and McCosker, 1991; Monterey Bay Aquarium, 2008b; Powell, 2001; David C. Powell, pers. comm. 2008, 2009).

1994: Sea World San Diego, San Diego, California

In 1994, two female great white sharks, approximately 155 cm long and 30 kg, were kept in captivity for 13 and 10 days respectively at Sea World San Diego. The two juvenile white sharks were collected off the southern California coast. Both sharks manifested severe hyperglycemia within 10 days of being placed into the 400,000 gallon exhibit at Sea World. On arrival the first shark, which was caught in a fisherman's net, was given an injection of an anti-inflammatory (dexamethasone sodium phosphate) and an antibiotic (amikacin sulfate). Over the next five days, the shark's swimming pattern appeared normal, but it did not show interest in any of the live fish swimming in the exhibit, which included spiny dogfish *Squalus acanthias* (Linnaeus, 1758), mackerel and bat rays *Myliobatis californica*. Moreover, the white shark did not show interest in any of the food offered to it, including salmon, rockfish, California scorpionfish *Scorpaena guttata* (Girard, 1854), halibut, mackerel, barracuda *Sphyraena* sp., squid, skate wings or starry flounder *Platichthys stellatus* (Pallas, 1787). In the hope of stimulating the shark's appetite, a second injection of dexamethasone was administered on day five. Over the next five days there was no clinical change, so a third injection of dexamethasone was administered, and the shark was force fed a 1 kg mixture of herring and mackerel. Within the next 3 days, the shark's swimming posture became abnormal and it was euthanized. On day 10, prior to the last injection of dexamethasone, the shark was hyperglycemic (blood glucose level of 754 mg/dl, whereas 100 mg/dl is normal for other shark species) with a serum insulin less than 2.5 mcU/ml. On day 13, the insulin level rose to an unbelievable level of 2961 mcU/ml.

Upon arrival of the second shark, which was caught on a long line, a blood sample was collected and an intramuscular injection of antibiotic (amikacin sulfate) was administered. All blood values were normal, including blood glucose and insulin levels. This shark quickly learned to negotiate the new environment, but, as with the first shark, it showed no interest in any of the live or offered fish. On day 10, the shark was released having shown neither an appetite nor clinical signs of deterioration. A blood sample taken on the day of release demonstrated hyperglycemia (blood glucose level of 353 mg/dl), but this shark showed no insulin response. The reason why the hyperglycemia occurred in these specimens remained unclear (Reidarson and McBain, 1994).

May 2000: San Pedro, California

On May 30, 2000, a 137 cm female great white shark was kept in captivity for more than three days. The shark was inadvertently caught in a gill net and taken to San Pedro, California. It was kept for husbandry research and satellite tagging. After capture, it was held in a bait tank with running seawater aboard the fishing boat for a minimum of five hours. Then it was transferred to a 6.1 m diameter x 1.8 m deep tank with running seawater. It is important to start these sharks swimming again after containment in a tank. This can be done by walking them or, more effectively, by facing and holding them in a strong current, which was done in this case. The shark started swimming continuously on its own within an hour of being released into the tank. It continued to regain swimming strength and the abrasions from capture and transport appeared to be healing by the end of three days in captivity. A whole freshly killed mackerel was suspended with a monofilament line 0.6 m below the surface 24 hours later. The food was missing the following morning. A shark tooth was discovered on the bottom directly below where the mackerel had been. There is the possibility that a seabird may have taken the mackerel, but this is unlikely because the tank was 90 percent covered with a tarp and the mackerel was submerged two feet under the surface. The shark was

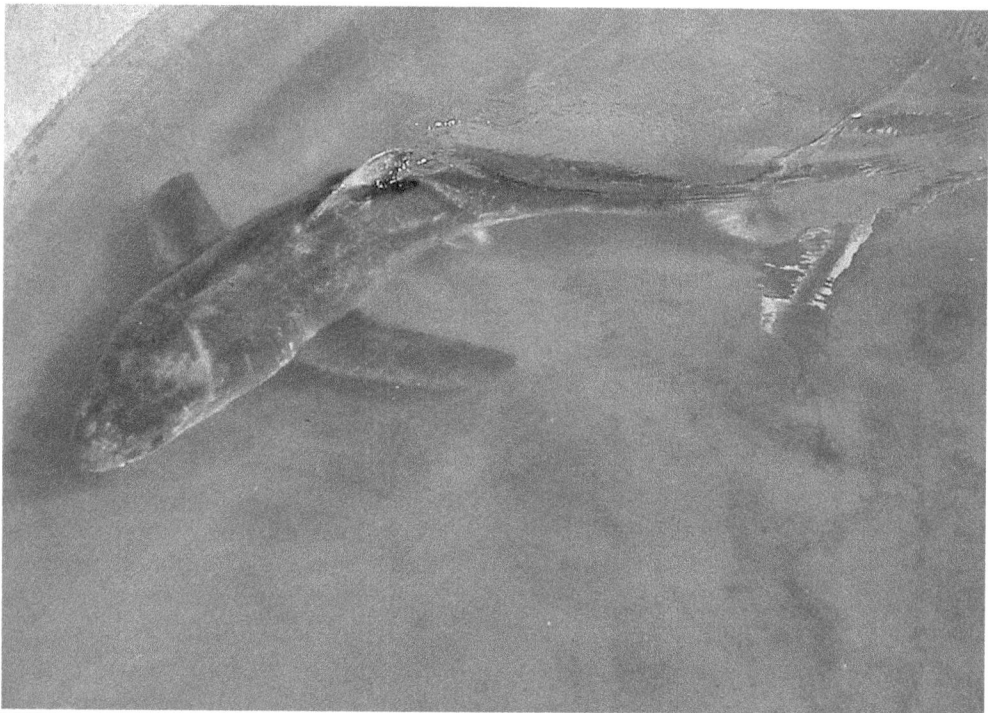

In May 2000, this 137 cm female great white shark was kept in captivity for more than three days (photograph courtesy Chuck Winkler).

released after 3⅓ days in captivity in the water offshore of Los Angeles Harbor with a pop-up archival satellite tag attached to it. The shark tag was recovered one month later in another fisherman's net in the same area where it was caught the first time and approximately 18 miles from where it was released. The shark apparently freed himself from the net but left the pop-up tag behind tangled in the net. The tag was found to contain a complete set of environmental data that clearly demonstrated that the shark had survived its capture and release (Henry Mollet's website, 2008; Chuck Winkler, pers. comm. 2009).

September 2004: Monterey Bay Aquarium, Monterey, California

The exhibit of white sharks is one part of the Monterey Bay Aquarium White Shark Conservation Research Project. Since 2002, the project has aimed to learn more about white sharks in the wild as well as to bring white sharks to Monterey for exhibit. The Monterey Bay Aquarium white shark project is a collaborative, multi-year study of white sharks off the California coast. The project has two primary goals, which include tagging and field studies of wild sharks, and the exhibition of captured white sharks. The Monterey Bay Aquarium staff is conducting field studies to tag juvenile and adult white sharks, tracking where they go in the ocean. The objective is to study how white sharks fit into the ocean ecosystem to help fisheries managers develop better ways to protect them. This work is a collaborative effort with their research partners at the following institutions and research programs: Stanford University, Tagging of Pacific Predators, California State University, Long Beach, University of California, Davis, Point Reyes Bird Observatory Conservation Science, and the Pelagic Shark Research Foundation. To date, scientists have tagged and tracked 18 juvenile white sharks and 143 adults. In general, the juvenile sharks have tended to remain in the coastal zone ranging from Northern California to Baja, Mexico. Adult sharks have traveled as far west as Hawaii.

In the opinion of the Monterey Bay Aquarium staff, exhibiting a young white shark allows them to contribute significantly to public understanding and protection of white sharks, an ecologically important and increasingly threatened species. Their ongoing efforts to study and possibly exhibit a young white shark began in 2002. The Monterey Bay Aquarium took a cautious and methodical approach to exhibiting a white shark. This approach, developed in consultation with a panel of independent shark experts (John C. Hewitt, John Rupp, Frank Murru and David C. Powell), was designed to minimize the stresses of collection, holding and transport (John O'Sullivan, pers. comm. 2008). It focused on "young of the year" animals, drawing on years of experience that found younger animals acclimate more readily than adults. Smaller sharks are also easier to handle and transport (Monterey Bay Aquarium, 2008b).

The Monterey Bay Aquarium used two methods to collect white sharks. Sometimes their husbandry staff collects young white sharks directly, by hook-and-line or a seine net. They also rely on commercial fishing crews in Southern California, who occasionally catch juvenile white sharks accidentally while fishing for halibut. They have asked crews to contact them if they capture a young white shark that is alive and healthy. They have a rapid response team standing by to work with any sharks that are caught by commercial fishermen. Team members assess the sharks' health and either transfer them to an ocean holding pen, or tag and release them to the wild. There are many unknowns with sharks obtained as bycatch from a commercial fishery. Usually, the time caught in the net is unknown, and to what degree their health might have been compromised as a result is also unknown. For that reason, they have much more confidence starting with a healthy animal collected by their own team (Monterey Bay Aquarium, 2008b).

Prior to transport, the aquarium deploys an ocean holding pen, similar to those used by commercial tuna ranchers. From 2002 to 2004, a pen that was 45 m × 15 m was leased. The Monterey Bay Aquarium later purchased its own pen that measured 45 m × 11 m (John O'Sullivan, pers. comm. 2008). This pen provides a controlled environment in which a juvenile shark has a chance to recover from the stress of being caught in fishing gear and can be observed for injury or illness, and trained to accept prepared food (e.g., salmon filets, mackerel, etc.). The aquarium staff has to confirm that the shark is feeding before an attempt is made to transport it to the Monterey Bay Aquarium. Captivity in the pen also allows the shark to get used to navigating in a confined environment, which eases the transition to an exhibit setting. A 3,000-gallon mobile life support transport vehicle is used to transport great white sharks from the ocean holding pen to the Monterey Bay Aquarium (John O'Sullivan, pers. comm. 2008).

The Monterey Bay Aquarium has a 1.2-million-gallon Outer Bay exhibit that was designed for pelagic animals and engineered to accommodate sharks. For example, they dampened (as much as possible) the electrical field interference created in the exhibit by the life-support equipment. The Outer Bay exhibit is home to Galapagos sharks, scalloped hammerhead sharks, bluefin tuna, yellowfin tuna, barracuda, ocean sunfish and other species (Monterey Bay Aquarium, 2008b).

The Monterey Bay Aquarium feeds the young white sharks wild-caught salmon, mackerel and sardines, supplemented with specially formulated vitamins. Albacore tuna has also been added to the menu. If a white shark has to be returned to the wild, the aquarium staff tags it prior to release. The Monterey Bay Aquarium uses both a "pop-up archival tag" (PAT tag) and a "smart position-only tag" (SPOT tag). The pop-up tags are attached externally to a shark, where they collect data on temperature, depth and light (used to estimate position). They store the data in a tiny computer and on a preprogrammed date, the tag releases from the shark and floats to the surface. The data is then sent via satellite back to the laboratory, where it can be analyzed. If the aquarium staff is able to collect the tag floating in the water, more information can be obtained. The SPOT tags provide near real-time information about where the sharks go, and send the data to researchers via satellite until the batteries expire. This tag is typically attached to the dorsal fin and uploads data to a satellite when the shark is on the surface.

On July 29, 2003, a juvenile white shark was accidentally caught by a commercial fisherman and brought to the borrowed ocean pen, where it began feeding after three days. It remained in the pen for six days before it was tagged and released. The pen had to be returned to its owner before aquarium staff felt ready to bring the shark to Monterey (Monterey Bay Aquarium, 2008b).

After this episode, five more sharks were transferred to the pen, first in September 2004 and again in August 2006, 2007, 2008 and 2009. Monterey Bay Aquarium succeeded in exhibiting these young white sharks in their Outer Bay exhibit for a short time and subsequently returned four of them to the wild.

The Monterey Bay Aquarium received their first young white shark on August 20, 2004, after it was caught inadvertently by commercial fishermen in Southern California. For more than three weeks it was held in the ocean pen, where it remained in good health; the shark navigated the pen well and began feeding. On September 14, it was transported to Monterey and placed directly in the 1.2-million-gallon Outer Bay exhibit. The next morning it fed, which was the first time a white shark had successfully taken food while on exhibit. This became the longest-ever exhibit of a white shark. During the shark's 198 days in this million-gallon exhibit, it was seen by nearly a million visitors (half of whom said they came specifically to see it), and it grew from a length of 152.4 cm (5 ft) and a weight of 28.1 kg (62 lb) to a length of 194.3 cm (6 ft 4.5 in) and a weight of 73.5 kg (162 lb) (Monterey Bay Aquarium, 2008a, 2008b). The white shark bit and killed two tope sharks *Galeorhinus galeus*, though it wasn't clear that it was hunting them. When the shark's behavior changed and it began actively hunting scalloped hammerhead sharks *Sphyrna lewini* in the exhibit, the aquarium staff decided to return it to the wild within four days (Manny Ezcurra, pers. comm. 2008). On March 31, 2005, it was fitted with an electronic tag and released in the waters just south of Monterey Bay. Its movements were tracked for 30 days. As programmed, on April 30, 2005, the tag popped free and was recovered off the Santa Barbara County coast by a scientist from Stanford University. The data stored on the tag showed that the shark had traveled nearly 200 miles south and had been diving to depths greater than 243.8 m (Monterey Bay Aquarium, 2008a, 2008b).

August 2006: Monterey Bay Aquarium, Monterey, California

The Monterey Bay Aquarium's husbandry collectors caught their second white shark in 2006. On August 17, 2006, a year-old male juvenile white shark was caught by aquarium collectors on hook-and-line tackle offshore in Santa Monica Bay, and transferred to the ocean pen. The male shark was held in the pen for nearly two weeks, and was then transferred to the Outer Bay exhibit on August 31. During its 137 days at the Aquarium, it grew from a length of 172.7 cm (5 ft 8 in) and a weight of 46.7 kg (103 lb) to its release size of 195.6 cm (6 ft 5 in) and a weight of 77.6 kg (171 lb). On January 16, 2007, it was fitted with an electronic tag and released in Monterey Bay. Its tag popped free 90 days later, off the tip of Cabo San Lucas in Baja, California, a journey covering more than 2,200 miles that took it up to 700 miles offshore and to depths more than 304.8 m below the surface (Monterey Bay Aquarium, 2008a, 2008b).

August 2007: Monterey Bay Aquarium, Monterey, California

The Monterey Bay Aquarium acquired a third white shark that was caught accidentally in a commercial sea bass net on August 4, 2007, in waters off Ventura, southern California, and transferred to the ocean pen. The shark was held in the pen for 24 days, then transferred to the Outer Bay exhibit on August 28, 2007. It was on exhibit for 162 days. On arrival, it was 144.8 cm (4 ft 9 in) long and weighed 30.6 kg (67.5 lb). It grew to 177.8 cm (5 ft 10 in) and 63.5 kg (140 lb) prior to release. The shark was released in Monterey Bay on February 5, 2008. It was fitted with two tracking tags. The SPOT tag reported realtime data on the shark's location. In the first 40 days the shark traveled to the southern tip of Baja, California, and then swam halfway up the Sea of Cortez before the tag battery stopped in June. The second tag, a Popup Archival Tag, popped free in June, and data from the retrieved tag should reveal detailed information on the depth and length of the shark's dives.

August 2008: Monterey Bay Aquarium, Monterey, California

The Monterey Bay Aquarium acquired a fourth white shark that was collected on August 16, 2008 in Santa Monica Bay by aquarium collectors, with the help of a commercial fisherman using a seine net. In the summer of 2008, four other young white sharks were brought to the pen and ultimately released because they didn't demonstrate normal swimming and feeding behavior. This female shark, however, held in the 4-million-gallon ocean pen off Malibu, was observed swimming comfortably and feeding in the pen several times. On August 27, 2008, 12 days after it was caught, the Monterey Bay Aquarium staff brought the shark to Monterey and placed her in the Outer

During its 137 days at the Monterey Bay Aquarium, Monterey, California, this male white shark grew from a length of 172.7 cm to its release size of 195.6 cm (photograph courtesy Monterey Bay Aquarium/Randy Wilder).

Bay exhibit. The 137.2 cm (4.5 ft) long, 25.2 kg (55.5 lb) female shark fed only once while at the Monterey Bay Aquarium. Consequently, the animal care staff decided to return it to the wild. Though it looked strong, it was important to put the shark back in the ocean while it was healthy. It remained on exhibit for 11 days before being tagged and released to the wild in offshore waters in the Santa Barbara Channel on September 7. Like the three other white sharks exhibited at the Monterey Bay Aquarium, it carried a tracking tag to document its movements after release. The PAT tag popped free on October 8, four months ahead of schedule, and was recovered near San Miguel Island in the Santa Barbara Channel on October 23. Data from the tag shows that the shark had remained in waters around the Channel Islands, where it was released on September 7, and that it was doing well in the wild (Monterey Bay Aquarium, 2008a, 2008b).

August 2009: Monterey Bay Aquarium, Monterey, California

The Monterey Bay Aquarium acquired a fifth white shark that was collected on August 12, 2009, near Malibu, California, with the help of a spotter plane and a commercial fishing crew using a purse seine net. The 160.0 cm (5 ft 3 in) long, 36.2 kg (79.8 lb) female shark remained in the ocean holding pen for almost two weeks while the Aquarium staff confirmed that it was feeding on exhibit. At the time of writing the last lines of this book, the shark seems to be adapting well and is feeding consistently, so the Aquarium's staff hopes to have it with them for a few months (Manny Ezcurra, pers. comm. 2009; Ken Peterson, pers. comm. 2009).

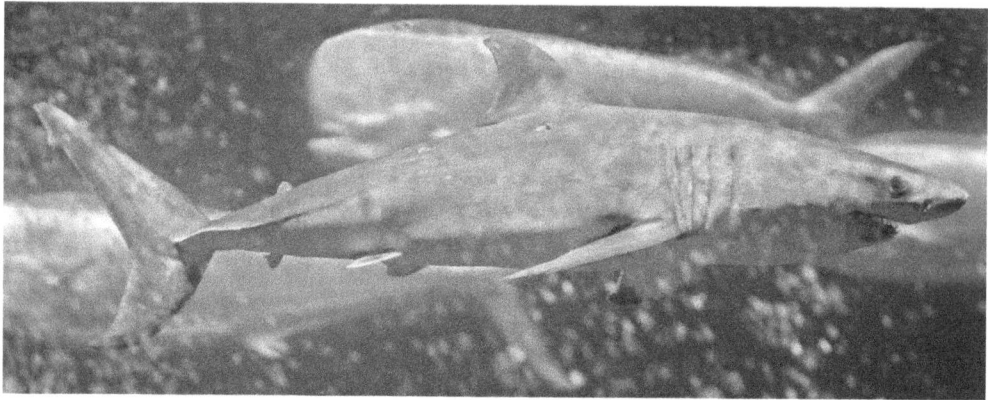

A 160.0 cm female great white shark was placed on exhibit August 26, 2009, at the Monterey Bay Aquarium, Monterey, California. It is the fifth great white shark to be placed on exhibit at the Aquarium (photograph courtesy Monterey Bay Aquarium/Randy Wilder).

The Monterey Bay Aquarium staff believes that the past successes demonstrate that it is possible to exhibit a white shark for an extended period of time, and to return it successfully to the wild (Monterey Bay Aquarium, 2008b).

Appendix I — Institutions Contacted

A total of 366 institutions have been contacted, including museums, aquariums and institutes of natural history, zoology, biology, anatomy and ichthyology across the entire United States. The institutions were not limited to those known for research on sharks. In general, as many as possible scientific institutions with both a strong or marginal orientation in marine zoology were contacted for this study. Following is the complete list of the institutions that have been contacted. They are ordered in alphabetical order by state, then by city and then by institution name:

Alabama

Anniston Museum of Natural History, Anniston
McWane Science Center, Birmingham
Sci-Quest — North Alabama Science Center, Huntsville
Alabama Museum of Natural History, University of Alabama, Tuscaloosa

Alaska

Alaska Museum of Natural History, Anchorage
Museum of the North, University of Alaska, Fairbanks
School of Fisheries and Ocean Sciences, University of Alaska Fairbanks, Fairbanks
Pratt Museum, Homer
Alaska State Museum, Juneau
Auke Bay Laboratories, Alaska Fisheries Science Center, National Marine Fisheries Service Juneau
Sheldon Jackson Museum, Sitka

Arizona

Museum of Northern Arizona, Flagstaff
Arizona Museum of Natural History, Mesa
Arizona-Sonora Desert Museum, Tucson
Flandrau: The University of Arizona Science Center, University of Arizona, Tucson
International Wildlife Museum, Tucson
Sonoran Sea Aquarium, Tucson

Arkansas

Museum of Discovery, Little Rock

California

Crab Cove Visitor Center, Alameda
Natural History Museum, Humboldt State University, Arcata
Lawrence Hall of Science, University of California, Berkeley
Berkeley Natural History Museums Consortium, Berkeley
Museum of Paleontology, University of California, Berkeley
Museum of Vertebrate Zoology, University of California, Berkeley
Bodega Marine Laboratory, University of California Davis, Bodega Bay
Chula Vista Nature Center, Chula Vista
The Raymond M. Alf Museum of Paleontology, Claremont
Museum of Jurassic Technology, Culver City
Fresno Metropolitan Museum of Art and Science, Fresno
Department of Biological Science, California State University Fullerton, Fullerton
California Surf Museum, Highway Oceanside
Ecology and Evolutionary Biology, University of California Irvine, Irvine
Orange County Natural History Museum, Laguna Niguel
Birch Aquarium at Scripps, La Jolla
Inter-American Tropical Tuna Commission, La Jolla

Appendix I — Institutions Contacted

Scripps Institution of Oceanography, University of California San Diego, La Jolla
Southwest Fisheries Science Center, National Marine Fisheries Service, La Jolla
Aquarium of the Pacific, Long Beach
California State University, Long Beach, Marine Laboratory, Long Beach
Natural History Museum of Los Angeles County, Los Angeles
Page Museum, Los Angeles
California Science Center, Los Angeles
Great Valley Museum of Natural History, Modesto
Monterey Bay Aquarium, Monterey
Museum of Natural History, Morro Bay State Park
Moss Landing Marine Laboratories, Moss Landing
Oakland Museum of California, Oakland
Pfleger Institute of Environmental Research, Oceanside
Hopkins Marine Station, Pacific Grove
Pacific Grove Museum of Natural History, Pacific Grove
Petaluma Wildlife Museum, Petaluma
Point Reyes Bird Observatory Conservation Science, Petaluma
Natural History Museum, Sierra College, Rocklin
California Department of Fish & Game, Sacramento
Conservation Science Institute, Santa Cruz
Pelagic Shark Research Foundation, Santa Cruz
San Diego Natural History Museum, San Diego
Sea World of California, San Diego
Aquarium of the Bay, San Francisco
California Academy of Sciences, San Francisco
Gulf of the Farallones National Marine Sanctuary, San Francisco
Randall Museum, San Francisco
Steinhart Aquarium, California Academy of Sciences, San Francisco
Coyote Point Museum, San Mateo
Cabrillo Marine Aquarium, San Pedro
Bowers Museum, Santa Ana
Discovery Science Center, Santa Ana
Ocean Futures Society, Santa Barbara
Santa Barbara Museum of Natural History, Santa Barbara
City of Santa Cruz Museum of Natural History, Santa Cruz
Institute of Marine Sciences, University of California, Santa Cruz
Santa Monica Pier Aquarium, Santa Monica
Six Flags Marine World, Vallejo
Shark Research Committee, Van Nuys
Lindsay Wildlife Museum, Walnut Creek

Colorado

Department of Environmental, Population, and Organismic Biology, University of Colorado, Boulder
Museum of Natural History, University of Colorado, Boulder
Denver Museum of Nature & Science, Denver
Ocean Journey, Denver
Discovery Science Center, Fort Collins
Morrison Natural History Museum, Morrison
Wildlife Experience, Parker

Connecticut

Bruce Museum, Greenwich
Department of Marine Sciences, University of Connecticut, Groton
Mystic Aquarium, Mystic
Peabody Museum of Natural History, Yale University, New Haven
The Maritime Aquarium at Norwalk, Norwalk
Stamford Museum & Nature Center, Stamford
Connecticut State Museum of Natural History, Storrs
Department of Biology, University of Hartford, West Hartford

Delaware

Delaware Museum of Natural History, Wilmington
Mid-Atlantic Fishery Management Council, Dover

Florida

BEECS Genetic Analysis Core, University of Florida, Alachua
South Florida Museum, Bradenton
Clearwater Marine Aquarium, Clearwater
Department of Biology, Cox Science Center, University of Miami, Coral Gables
Oceanographic Center, Nova Southeastern University, Dania Beach

Museum of Arts and Sciences, Daytona Beach
Guy Harvey Research Institute, Oceanographic Center, Nova Southeastern University, Fort Lauderdale
Department of Fisheries and Aquatic Sciences, University of Florida, Gainesville
Florida Museum of Natural History, University of Florida, Gainesville
Jaws International, Key Biscayne
Key West Aquarium, Key West
Miami Science Museum, Miami
Miami Seaquarium, Miami
Rosenstiel School of Marine and Atmospheric Science, University of Miami, Miami
Department of Biology, University of Central Florida, Orlando
Orlando Science Center, Orlando
Ripley Aquariums, Orlando
Sea World Orlando, Orlando
Panama City Laboratory, National Marine Fisheries Service, Panama City
Mote Aquarium, Sarasota
Mote Marine Laboratory, Sarasota
Marineland, St. Augustine
Museum of the Islands, St. James City
Fish and Wildlife Research Institute, St. Petersburg
NOAA Fisheries Service, Southeast Regional Office, Saint Petersburg
Coastal & Marine Laboratory, Florida State University, St. Teresa
Tallahassee Museum, Tallahassee
Florida Aquarium, Tampa
Department of Biology, University of South Florida, Tampa

Georgia

Georgia Museum of Natural History, University of Georgia, Athens
Fernbank Museum of Natural History, Atlanta
Georgia Aquarium, Atlanta
Museum of Arts and Sciences, Macon
Georgia Southern Museum, Statesboro

Hawaii

Bernice Bishop Museum, Honolulu
Honolulu Community College Dinosaur Exhibit, University of Hawaii, Honolulu
Department of Zoology, University of Hawaii at Manoa, Honolulu
Waikiki Aquarium, University of Hawaii at Manoa, Honolulu
Hawaii Institute of Marine Biology, Kaneohe
Koke'e Natural History Museum, Kauai
Lana'i Culture & Heritage Center, Lana'i City
Maui Ocean Center, Wailuku

Idaho

Orma J. Smith Museum of Natural History, College of Idaho, Caldwell
Idaho Museum of Natural History, Pocatello

Illinois

Illinois Natural History Survey, Champaign
Department of Biological Sciences, DePaul University, Chicago
Department of Biological Sciences, University of Illinois at Chicago, Chicago
Field Museum, Chicago
John G. Shedd Aquarium, Chicago
Museum of Science and Industry, Chicago
Peggy Notebaert Nature Museum, Chicago Academy of Sciences, Chicago
Elgin Public Museum, Elgin
Jurica Nature Museum, Benedictine University, Lisle
Lakeview Museum of Arts and Sciences, Peoria
Burpee Museum of Natural History, Rockford
Illinois State Museum, Springfield
Midwest Museum of Natural History, Sycamore

Indiana

Wonderlab, The Museum of Science, Health and Technology, Bloomington
Evansville Museum, Evansville
Sumner B. Sheets Museum of Wildlife and Marine Exhibits, Huntington
Indiana State Museum, Indianapolis
Indiana University School of Medicine, Notre Dame
Joseph Moore Museum, Earlham College, Richmond

Iowa

Putnam Museum, Davenport
Museum of Natural History, University of Iowa, Iowa City

Kansas

Johnston Geology Museum, Emporia State University, Emporia
Schmidt Museum of Natural History, Emporia State University, Emporia
Sternberg Museum of Natural History, Fort Hays State University, Hays
Kansas University Natural History Museum, Kansas University, Lawrence
Fick Fossil and History Museum, Oakley

Kentucky

Louisville Science Center, Louisville
Newport Aquarium, Newport
Owensboro Area Museum of Science and History, Owensboro
Cumberland Inn & Museum, Cumberland College, Williamsburg

Louisiana

Marine Fisheries Division, Louisiana Department of Wildlife & Fisheries, Baton Rouge
Louisiana Museum of Natural History, Louisiana State University, Baton Rouge
Lafayette Natural History Museum and Planetarium, Lafayette
Museum of Natural History, University of Louisiana at Monroe, Monroe
Audubon Aquarium of the Americas, New Orleans

Maine

Maine State Museum, Augusta
George B. Dorr Museum of Natural History, College of the Atlantic, Bar Harbor
Nylander Museum, Caribou
Wilson Museum, Castine
L.C. Bates Museum, Good Will-Hinckley, Hinckley
Maine Aquarium, Kennebunkport
Gulf of Maine Aquarium, Portland
Northern Maine Museum of Science, University of Maine, Presque Isle

Maryland

Maryland Science Center, Baltimore
National Aquarium in Baltimore, Baltimore
Natural History Society of Maryland, Baltimore
University of Maryland School of Medicine, Baltimore
NOAA Fisheries Service, Silver Spring
Calvert Marine Museum, Solomons

Massachusetts

Division of Marine Fisheries, Massachusetts Department of Fish and Game, Boston
Massachusetts Museum of Natural History, University of Massachusetts, Amherst
Boston University Marine Program, Boston University, Boston
Museum of Science, Boston
New England Aquarium, Boston
Cape Cod Museum of Natural History, Brewster
Museum of Comparative Zoology, Harvard University, Cambridge
Sustainable Fisheries Division, National Marine Fisheries Service, Gloucester
Marion Natural History Museum, Marion
Natural Science Museum in Hinchman House, Nantucket
Biology Department, University of Massachusetts Dartmouth, North Dartmouth
Berkshire Museum, Pittsfield
Springfield Science Museum, Springfield
Woods Hole Oceanographic Institution, Woods Hole
EcoTarium, Worcester

Michigan

Department of Biology, Albion College, Albion
Exhibit Museum of Natural History, University of Michigan, Ann Arbor
Museum of Zoology, University of Michigan, Ann Arbor
Kingman Museum, Battle Creek
Card Wildlife Education Center and Wildlife Museum, Ferris State University, Big Rapids
Cranbrook Institute of Science, Bloomfield Hills
Michigan State University Museum, East Lansing
Museum of Cultural and Natural History, Central Michigan University, Mt. Pleasant

Minnesota

Underwater Adventures, Bloomington
Bell Museum of Natural History, University of Minnesota, Minneapolis
Minnesota Zoo, Apple Valley
Science Museum of Minnesota, St. Paul

Mississippi

Mississippi Museum of Natural Science, Jackson
Center for Fisheries Research and Development, University of Southern Mississippi, Ocean Springs
Mississippi Laboratories, Southeast Fisheries Science Center, National Marine Fisheries Service, Pascagoula

Missouri

Kansas City Museum, Kansas City
St. Louis Science Center, St. Louis

Montana

Phillips County Museum, Malta
Philip L. Wright Zoological Museum, The University of Montana, Missoula

Nebraska

WyoBraska Wildlife Museum, Gering
Hastings Museum of Natural and Cultural History, Hastings
University of Nebraska State Museum, Lincoln
Omaha Henry Doorly Zoo, Omaha

Nevada

Las Vegas Natural History Museum, Las Vegas
Marjorie Barrick Museum of Natural History, Las Vegas
Shark Reef, Las Vegas

New Hampshire

Little Nature Museum at Gould Hill Orchards, Contoocook

New Jersey

Adventure Aquarium, Camden
Morris Museum, Morristown
Newark Museum, Newark
Bergen Museum of Art & Science, Paramus
Museum of Natural History, Princeton University, Princeton
Shark Research Institute, Princeton
New Jersey State Museum, Trenton

New Mexico

Albuquerque Aquarium, Albuquerque
Museum of Southwestern Biology, University of New Mexico, Albuquerque
New Mexico Museum of Natural History and Science, Albuquerque
Museum of Natural History, Las Cruces

New York

New York State Museum, University of the State of New York, Albany
Roberson Museum and Science Center, Binghamton
Buffalo Museum of Science, Buffalo
Vanderbilt Museum, Centerport
Pember Museum of Natural History, Granville
Department of Biology, Hofstra University, Hempstead
Paleontological Research Institution, Ithaca
American Museum of Natural History, New York City
New York Aquarium, The Wildlife Conservation Society, New York City
South Street Seaport Museum, New York City
Atlantis Marine World Aquarium, Riverhead
Rochester Museum & Science Center, Rochester
Charles Dickert Wildlife Collection, Saranac Lake
Staten Island Museum, Staten Island
Museum of Long Island Natural Science, Stony Brook University, Stony Brook
Wild Center, Tupper Lake

North Carolina

The North Carolina Aquarium at Pine Knoll Shores, Atlantic Beach

North Carolina Maritime Museum, Beaufort
Department of Marine Sciences, University of North Carolina, Chapel Hill
North Carolina Museum of Life and Science, Durham
Rankin Museum of American and Natural History, Ellerbe
Schiele Museum of Natural History, Gastonia
The North Carolina Aquarium at Fort Fisher, Kure Beach
North Carolina Aquarium on Roanoke Island, Manteo
Center for Marine Sciences and Technology, North Carolina State University, Morehead City
Museum of Coastal Carolina, Ocean Isle Beach
Onslow County Museum, Richlands
North Carolina State Museum of Natural Sciences, Raleigh
Cape Fear Museum of History and Science, Wilmington

North Dakota

North Dakota Heritage Center, Bismarck

Ohio

Cincinnati Museum Center, Cincinnati
Cleveland Museum of Natural History, Cleveland
Museum of Biological Diversity, Columbus
Columbus Zoo and Aquarium, Powell

Oklahoma

Oklahoma Biological Survey, University of Oklahoma, Norman
Sam Noble Oklahoma Museum of Natural History, University of Oklahoma, Norman

Oregon

Museum of Natural and Cultural History, University of Oregon, Eugene
Oregon Coast Aquarium, Newport
Oregon Museum of Science and Industry, Portland

Pennsylvania

State Museum of Pennsylvania, Harrisburg
North Museum of Natural History & Science, Lancaster
Academy of Natural Sciences, Philadelphia
Franklin Institute Science Museum, Philadelphia
Wagner Free Institute of Science, Philadelphia
Carnegie Museum of Natural History, Pittsburgh
Pittsburgh Zoo & PPG Aquarium, Pittsburgh
Reading Public Museum, Reading
Everhart Museum, Scranton

Rhode Island

Department of Biological Sciences, University of Rhode Island, Kingston
Department of Environmental and Natural Resource Economics, Kingston Coastal Institute, University of Rhode Island, Kingston
Apex Predators Investigation, National Marine Fisheries Service, Narragansett
Museum of Natural History and Cormack Planetarium, Providence

South Carolina

Charleston Museum, Charleston
National Ocean Service Marine Forensics Branch, Center for Coastal Environmental Health and Biomolecular Research, Charleston
Marine Resources Research Institute, South Carolina Department of Natural Resources, Charleston
South Carolina Aquarium, Charleston
Bob Campbell Geology Museum, Clemson University, Clemson
McKissick Museum, University of South Carolina, Columbia
South Carolina State Museum, Columbia
Horry County Museum, Conway
Florence Museum of Art, Science and History, Florence
Coastal Discovery Museum, Hilton Head Island
Ripley's Aquarium, Myrtle Beach
Colleton Museum, Walterboro

Appendix I — Institutions Contacted

South Dakota

Great Plains Zoo and Delbridge Museum of Natural History, Sioux Falls
Washington Pavilion of Arts and Science, Sioux Falls
W.H. Over Museum of Natural and Cultural History, Vermillion

Tennessee

Tennessee Aquarium, Chattanooga
Frank H. McClung Museum, University of Tennessee, Knoxville

Texas

Amarillo College Natural History Museum, Amarillo
Texas Natural Science Center, University of Texas at Austin, Austin
Brazos Valley Museum of Natural History, Bryan
Texas State Aquarium, Corpus Christi
Department of Wildlife and Fisheries Sciences, Texas A&M University, College Station
Dallas World Aquarium, Dallas
Museum of Nature & Science, Dallas
El Campo Museum of Natural History, El Campo
Centennial Museum, University of Texas, El Paso
Fort Worth Museum of Science and History, Fort Worth
Houston Museum of Natural Science, Houston
Brazosport Museum of Natural Science, Lake Jackson
International Museum of Art & Science, McAllen
Heard Natural Science Museum, McKinney
Museum of the Gulf Coast, Port Arthur
Buckhorn Saloon & Museum, San Antonio
San Antonio Zoological Gardens and Aquarium, San Antonio
Sea World San Antonio, San Antonio
Woodlands Science & Art Center, The Woodlands
Mayborn Museum, Baylor University, Waco

Virgin Islands

Coral World, St. Thomas

Utah

John Hutchings Museum of Natural History, Lehi
College of Eastern Utah Prehistoric Museum, Price
Weber State University Museum of Natural Science, Ogden
Paunsagaunt Wildlife Museum, Panguitch
Monte L. Bean Life Science Museum, Brigham Young University, Provo
Rosenbruch Wildlife Museum, St. George
Utah Museum of Natural History, University of Utah, Salt Lake City
Utah Field House of Natural History State Park Museum, Vernal

Vermont

Nature Museum, Grafton
Southern Vermont Natural History Museum, Marlboro
Montshire Museum of Science, Norwich
Fairbanks Museum & Planetarium, St. Johnsbury

Virginia

Virginia Institute of Marine Science, College of William and Mary, Gloucester Point
Hostetter Museum of Natural History, Eastern Mennonite University, Harrisonburg
Virginia Museum of Natural History, Martinsville
Virginia Living Museum, Newport News
Science Museum of Virginia, Richmond
Virginia Aquarium & Marine Science Center, Virginia Beach
U.S. Fish & Wildlife Service, Arlington

Washington

Whale Museum, Friday Harbor
Burke Museum, University of Washington, Seattle

Pacific Science Center, Seattle
Seattle Aquarium, Seattle
University of Washington Shark Research Lab, School of Aquatic and Fishery Sciences, Seattle
University of Washington Fish Collection, Seattle
Department of Biology, University of Puget Sound, Tacoma
Point Defiance Zoo & Aquarium, Tacoma
Slater Museum of Natural History, University of Puget Sound, Tacoma

Washington, D.C.

National Geographic Museum at Explorer Hall
Smithsonian Institution National Museum of Natural History

Wisconsin

Cable Natural History Museum, Cable
Geology Museum, University of Wisconsin—Madison, Madison
Milwaukee Public Museum, Milwaukee
Museum of Natural History, University of Wisconsin-Stevens Point, Stevens Point

Wyoming

National Museum of Wildlife Art, Jackson Hole
Geological Museum, University of Wyoming, Laramie
Natural History Museum, Western Wyoming College, Rock Springs
Wyoming Dinosaur Center, Thermopolis, Wyoming.

Appendix II — Reporting Specimens of Great White Sharks

Everyone who wishes to communicate with the author regarding information on white shark specimens preserved in museums not represented in this book, from the United States or from any part of the world, can contact him at the address below. For this reason, a special form is included. The information that is filled in must be as accurate as possible. However, it is typical that some entries are left blank, because it is very rare that all information is available. All data, even though it may be very little, may be very important.

The way to collect the data is as follows. Total length has to be measured in a straight line, from the tip of the snout to the tip of the upper lobe of the caudal fin, with the caudal fin in the natural position (TLn) or with the caudal fin in the depressed position (TOT) (see page 18, "Measurement and Estimation of Size"). If the reported length is just an estimate, this must be specified. If the jaws can be examined, the dried upper jaw perimeter (DUJP) has to be measured and, if possible, the enamel height of the largest upper tooth (UAE1 and UAE2) should be measured (see page 18, "Measurement and Estimation of Size"). The weight must be of the whole specimen, and if the animal has been gutted, it must be noted. If the reported weight is just an estimate, this must be specified. The sex, as already described in the chapter about reproduction, is easily recognized by looking at the ventral side of the animal. The males have the claspers, the reproductive organs, which are two cylindrical appendices that start at the base of the pelvic fins. With regards to conserving anatomical parts, it is best that the jaws are preserved complete, or if not possible, some of the anterior upper and lower teeth. The other anatomical parts and the embryos should be frozen or preserved in alcohol or formalin. Where photos are concerned, it is advisable to take one side photo of the entire subject. Other photos should include any unusual details, such as scars or parasites. In the photo, try to include a meter measuring stick or, if not possible, something else that can indicate the size dimensions. If dealing with a very large shark, and a measuring tool is not available, a human with noted height in cm, would be a very good reference. The photo should be taken from the side, with reference to the animal, of the complete animal with the size reference next to it. The photographer should stand a few meters away.

The form can be sent by normal or electronic post together with the photographs, to the following address:

>Dott. Alessandro De Maddalena
>Italian Ichthyological Society
>via L. Ariosto 4, 20145 Milano, Italy
>E-mail: a-demaddalena@tiscali.it

ITALIAN ICHTHYOLOGICAL SOCIETY

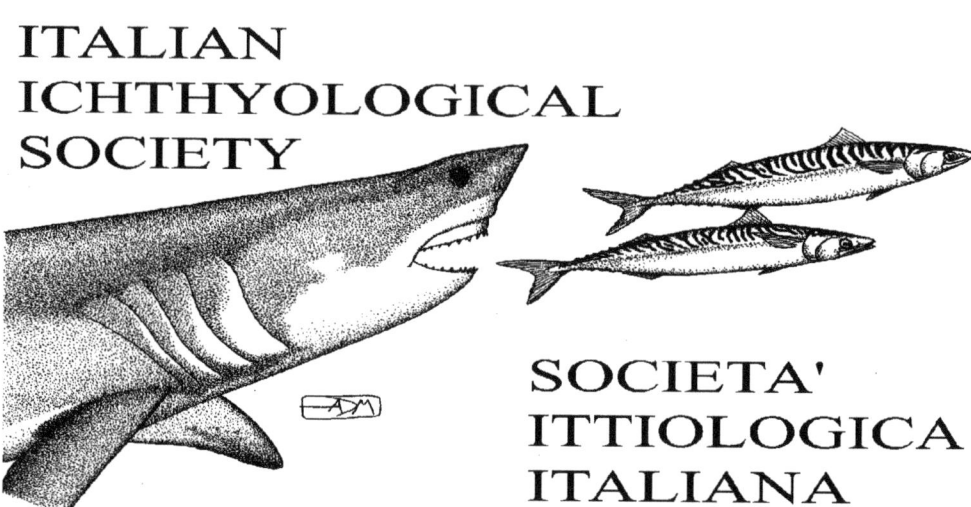

SOCIETA' ITTIOLOGICA ITALIANA

Preserved Great White Shark Materials Report Form

Species: great white shark *Carcharodon carcharias* (Linnaeus, 1758)

Institution name:

Institution address (street, city, state, ZIP, Country, email):

Date of acquisition:

Collector:

Determiner:

Storage or exhibit location:

Catalogue number:

Type of material:

Preservation method:

Date of capture:

Time of capture:

Place of capture:

Position (latitude and longitude):

Distance from shore:

Depth of sea:

Depth of capture:

Atmospheric conditions:

Sea conditions:

Total length (in a straight line, from the tip of the snout to the tip of the upper lobe of the caudal fin; specify if TLn or TOT):

Enamel height of the largest upper tooth (UAE1 and UAE2):

Dried upper jaw perimeter (DUJP):

Weight (specify if intact or gutted):

Sex:

Stomach contents:

Specimen activity at the time of capture:

Presence of other animals in the area at time of capture:

Activity of witnesses at time of encounter:

If it is a pregnant female, specify:

   Number of embryos:

   Total length of each embryo:

   Weight of each embryo:

   Sex of each embryo:

Photographic evidence (if possible, attach):

Specify if data and photos supplied can be published:

Remarks:

References:

Details of compiler

Name and Surname:

Address:

Telephone:

E-mail:

This report can be sent to:

        Dott. Alessandro De Maddalena
        Italian Ichthyological Society
        via L. Ariosto 4, 20145 Milano, Italy
        E-mail: a-demaddalena@tiscali.it

Bibliography

Ainley, D. G., Henderson, R. P., Huber, H. R., Boekelheide, R. J., Allen, S. G. and McElroy, T. L. (1985). Dynamics of white shark / pinniped interactions in the Gulf of the Farallones. *Memoirs of the Southern California Academy Of Sciences*, 9, 109–122.

Ainley, D. G., Strong, C. S., Huber, H. R., Lewis, T. J. and Morrell, S. H. (1981). Predation by sharks on pinnipeds at the Farallon Islands. *Fishery Bulletin*, 78, 941–945.

Ames, J. A., Geibel, J. G., Wendell, F. E. and Pattison, C. A. (1996). White shark-inflicted wounds of sea otters in California, 1968–1992. In A. P. Klimley and D. G. Ainley (Eds.), *Great white sharks: The biology of* Carcharodon carcharias (pp. 309–316). San Diego: Academic Press.

Anderson, S. D. and Goldman, K. J (1996). Photographic evidence of white shark movements in California waters. *California Fish and Game*, 82, 182–186.

Anonymous (1961a): Huge shark caught, near death here. *Honolulu Star Bulletin*, March 9, 1961: 1–2.

Anonymous (1961b): Monster. *Honolulu Advertiser*, March 9, 1961: 8.

Applegate, S. P. and Espinosa-Arrubarrena, L. (1996). The fossil history of *Carcharodon* and its possible ancestor, *Cretolamna*: a study in tooth identification. In A. P. Klimley and D. G. Ainley (Eds.), *Great white sharks: The biology of* Carcharodon carcharias (pp. 19–36). San Diego: Academic Press.

Arnold, P. W. (1972). Predation on harbour porpoise, *Phocoena phocoena*, by a white shark, *Carcharodon carcharias*. *Journal Fisheries Research Board of Canada*, 29, (8), 1213–1214.

Barrull, J. and Mate, I. (2001). Presence of the great white shark *Carcharodon carcharias* (Linnaeus, 1758) in the Catalonian Sea (NW Mediterranean): review and discussion of records, and notes about its ecology. *Annales, Series historia naturalis*, 11, (1), 3–12.

_____. (2002). *Tiburones del Mediterráneo*. Arenys de Mar: Libreria El Set-ciències.

Benz, G. W., Mollet, H. F., Ebert, D. A., Davies, C. R. and Van Sommeran, S.R. (2003). Five species of parasitic copepods (Siphonostomatoida: Pandaridae) from the body surface of a white shark captured in Morro Bay, California. *Pacific Science*, 57, 39–43.

Bigelow, H. B. and Schroeder, W. C. (1948). *Sharks. Fishes of the Western North Atlantic. Part one: lancelets, ciclostomes, sharks*. Memoir Sears Foundation for Marine Research. New Haven: Yale University, 53–576.

Black, C. (2008). *White Pointer South: the Tasmanian white shark chronicles*. Hobart, Australia: Ragged Tooth Productions.

Bonfil, R., Meyer, M., Scholl, M. C., Johnson, R., O'Brien, S., Oosthuizen, H., Swanson, S., Koetze, D. and Peterson, M. (2005). Transoceanic migration, spatial dynamics, and population linkages of white sharks. *Science*, 310, 100–103.

Boston Society of Natural History (1851). *Proceedings of the Boston Society of Natural History*, 3, 1–396.

Brian, A. (1906). *Copepodi parassiti dei pesci d'Italia*. Genova: Istituto Sordomuti.

Brown B.R., Hughes M.E. and Russo, C. (2005). Infrastructure in the electric sense: Admittance data from shark hydrogels. *Journal of Comparative Physiology A*, 191, 115–123.

Bruner, J. C. (1998). Tooth replacement rate of *Carcharodon carcharias* (Linnaeus, 1758). In: *AES 14th annual meeting. Program and Abstracts*, (98). Guelph, Ontario, Canada.

Burgess, G. H. and Callahan, M. (1996). Worldwide patterns of white shark attacks on humans. In A.P. Klimley and D. G. Ainley (Eds.), *Great white sharks: The biology of* Carcharodon carcharias (pp. 457–469). San Diego: Academic Press.

Cadenat, J. and Blache, J. (1981). *Requins de Mediterranée et d'Atlantique (plus particulièrement de la Côte Occidentale d'Afrique)*. Paris,

France: Editions de l'Office de la Recherche Scientifique et Technique Outre-mer.

Cailliet, G. M. (1990). Elasmobranch age determination and verification; an updated review. In H. L. Pratt Jr., S. H. Gruber, and T. Taniuchi (Eds.), Elasmobranchs as Living Resources: Advances in the Biology, Ecology, Systematics and the Status of the Fisheries (pp. 157–165). *NOAA Technical Report NMFS*, 90.

———. (1996). An evaluation of methodologies to study the population biology of white sharks. In A. P. Klimley and D. G. Ainley (Eds.), *Great white sharks: The biology of* Carcharodon carcharias (415–416). San Diego: Academic Press, San Diego.

Cailliet, G. M., Martin, L. K., Kusher, D., Wolf, P. and Weldon, B. A. (1983). Techniques for enhancing vertebral bands in age estimation of California elasmobranchs. In E. D. Prince and L. M. Pulos (Eds.), Proceedings of the international workshop on age determination of oceanic pelagic fishes: tunas, billfishes and sharks (pp. 157–165). *NOAA Technical Report NMFS*, 8.

Cappo, M. (1988). Size and age of the white pointer shark *Carcharodon carcharias* (Linnaeus, 1758); was Peter Riseley's white pointer a world record? *Safish*, 13, (1), 11–13.

Carey, F. G., Casey, J. G., Pratt, H. L., Urquhart, D. and McCosker, J. E. (1985). Temperature, heat production and heat exchange in lamnid sharks. *Memoirs of the Southern California Academy of Sciences*, 9, 92–108.

Carey, F. G., Kanwisher, J. W., Brazier, O., Gabrielson, G., Casey, J. G. and Pratt, H. L., Jr. (1982). Temperature and activities of a white shark, *Carcharodon carcharias*. *Copeia*, 1982, (2), 254–260.

Carey, F. G. and Scharold, J. V. (1990). Movements of blue sharks (*Prionace glauca*) in depth and course. *Marine Biology*, 106, 329–342.

Carrada, G. (1998). Lo squalo bianco è di casa nei mari italiani. *National Geographic Italia*, 2, (6), 165.

Casey, J. G. and Pratt, H. L. (1985). Distribution of the white shark, *Carcharodon carcharias*, in the western North Atlantic. *Memoirs of the Southern California Academy of Sciences*, 9, 2–14.

———. (1986). White sharks in the western North Atlantic. *Maritimes*, November 1986: 4–6.

Castro, J. (1983). *The sharks of North American waters*. College Station: Texas A&M University Press.

Celona, A., De Maddalena, A. and Comparetto, G. (2006). Evidence of a predatory attack on a bottlenose dolphin *Tursiops truncatus* by a great white shark *Carcharodon carcharias* from the Mediterranean Sea. *Annales, Series historia naturalis*, 16, (2), 159–164.

Celona, A., Donato, N. and De Maddalena, A. (2001). In relation to the captures of a great white shark *Carcharodon carcharias* (Linnaeus, 1758) and a shortfin mako, *Isurus oxyrinchus* (Rafinesque, 1809) in the Messina Strait. *Annales, Series historia naturalis*, 11, (1), 13–16.

Cigala Fulgosi, F. (1990). Predation (or possible scavenging) by a great white shark on an extinct species of bottlenosed dolphin in the Italian Pliocene. *Tertiary Research*, 12, (1), 17–36.

Clarke, S. C., McAllister, M. K., Milner-Gulland, E. J., Kirkwood, G. P., Michielsens, C. G., Agnew, D. J., Pikitch, E. K., Nakano, H. and Shivji, M. S. (2006). Global estimates of shark catches using trade records from commercial markets. *Ecology Letters*, 9, (10), 1115–26.

Cliff, G., Compagno, L. J. V., Smale, M. J., Van Der Elst, R. P. and Wintner, S. P. (2000). First records of white sharks, *Carcharodon carcharias*, from Mauritius, Zanzibar, Madagascar and Kenya. *South African Journal of Science*, 96, 365–367.

Cliff, G., Dudley, S. F. J. and Davis, B. (1989). Sharks caught in the protective gill nets off Natal, South Africa. 2. The great white shark *Carcharodon carcharias* (Linnaeus, 1758). *South African Journal of Marine Sciences*, 8, 131–144.

Cliff, G., Dudley, S. F. J. and Jury, M. R. (1996). Catches of white sharks in KwaZulu-Natal, South Africa and environmental influences. In A. P. Klimley and D. G. Ainley (Eds.), *Great white sharks: The biology of* Carcharodon carcharias (pp. 351–361). San Diego: Academic Press.

Cliff, G., Van Der Elst, R. P., Witthuhn, T. K. and Bullen, E. M. (1996). First estimates of mortality and population size of white sharks on the South African coast. In A. P. Klimley and D. G. Ainley (Eds.), *Great white sharks: The biology of* Carcharodon carcharias (pp. 393–400). San Diego: Academic Press.

Coles, R. J. (1919). The large sharks of Cape Lookout, North Carolina. The white shark or maneater, tiger shark and hammerhead. *Copeia*, 69, 34–43.

Collier, R. (2003). *Shark attacks of the twentieth century from the Pacific coast of North America*. Chatsworth: Scientia Publishing, LLC.

Collier, R. S., Marks, M. and Warner, R. W. (1996). White shark attacks on inanimate objects along the Pacific coast of North America. In A.P. Klimley and D. G. Ainley (Eds.), *Great white sharks: The biology of* Carcharodon carcharias (pp. 217–221). San Diego: Academic Press.

Compagno, L. J. V. (1984). Sharks of the World. An annotated and illustrated catalogue of sharks species known to date. Parts 1 and 2. *FAO Species Catalogue, Vol.4, FAO Fisheries Synopsis*, 125, 1–655.

_____. (1999). Systematics and body form. In W. C. Hamlett (Ed.), *Sharks, skates, and rays: the biology of elasmobranch fishes* (pp. 1–42). Baltimore: Johns Hopkins University Press, Baltimore.

_____. (2001). *Sharks of the World. Volume 2. FAO Species Catalogue for Fishery Purposes, No. 1 Vol. 2.*

Conrath, C. L. (2005). Reproductive biology. In J. A. Musick and R. Bonfil (Eds), *Management techniques for elasmobranch fisheries* (pp. 103–126). *FAO Fisheries Technical Paper*, 474.

Council of the European Union (2006). Council Regulation (EC) No 1782/2006 of 20 November 2006 amending Regulations (EC) No 51/2006 and (EC) No 2270/2004, as regards fishing opportunities and associated conditions for certain fish stocks. *Official Journal of the European Union*, L 345 EN, 8/12/2006: 10–23.

Cousteau, J-M. and Richards, M. (1992). *Cousteau's great white shark*. New York: Harry N. Abrams.

Curtis, T., Menard, K. and Laroche, K. (2000). *A firsthand account of three white sharks scavenging a whale carcass*. Florida Museum of Natural History website: <http://www.flmnh.ufl.edu/fish/sharks/White/tobeycurtis.htm>

Dailey, M. D. and Vogelbein, W. (1990). *Clistobothrium carcharodoni* gen. et sp. n. (Cestoda: Tetraphyllidea) from the spiral valve of the great white shark (*Carcharodon carcharias*). *Journal of the Helminthological Society of Washington*, 57, 108–112.

Dehart, A. (2004). Species Selection and Compatibility. In M. Smith, D. Warmolts, D. Thoney and R. Hueter (Eds.), *The elasmobranch husbandry manual: captive care of sharks, rays and their relatives* (pp. 15–23). Columbus: Ohio Biological Survey, Inc.

De Maddalena, A. (1997). *Osservazioni sulla presenza e distribuzione di Carcharodon carcharias (Linnaeus, 1758) nel Mare Mediterraneo: segnalazioni e reperti museali*. Milano: Tesi di Laurea. Università degli Studi di Milano, Facoltà di Scienze Matematiche, Fisiche e Naturali, Milano.

_____. (1998). Il più grande esemplare italiano di squalo bianco, *Carcharodon carcharias* (Linnaeus, 1758) individuato nei reperti conservati presso il Museo di Anatomia Comparata dell'Università "La Sapienza" di Roma. *Museologia Scientifica*, 15, (2), 195-198.

_____. (2000a). Historical and contemporary presence of the great white shark *Carcharodon carcharias* (Linnaeus, 1758), in the Northern and Central Adriatic Sea. *Annales, Series historia naturalis*, 10, (1), 3–18.

_____. (2000b). Sui reperti di 28 esemplari di squalo bianco, *Carcharodon carcharias* (Linnaeus, 1758), conservati in musei italiani. *Annali del Museo Civico di Storia Naturale "G. Doria," Genova*, 93, 565–605.

_____. (2001b). Nel mar Mediterraneo non c'è solo il bianco. *Quark*, 4, Giugno 2001: 64.

_____. (2002). *Lo squalo bianco nei mari d'Italia*. Ireco, Formello.

_____. (2006a). The great white shark, *Carcharodon carcharias* (Linnaeus, 1758) of the Settala Museum in Milan. *Bollettino del Museo civico di Storia naturale di Venezia*, 57, 149-154.

_____. (2006b). A catalogue of great white sharks *Carcharodon carcharias* (Linnaeus, 1758) preserved in European museums. *Časopis Národního muzea, Řada přírodovědná (Journal of the National Museum, Natural History Series)*, 175, (3-4), 109-125.

_____. (2007). *Great white sharks preserved in European museums*. Newcastle upon Tyne: Cambridge Scholars Publishing.

_____. (2008). *Sharks—The perfect predators*. Houghton: Jacana Media.

_____. (2009). Lo squalo bianco nel Mediterraneo. *Rivista Marittima, Supplemento*. In press.

De Maddalena, A. and Baensch, H. (2008). *Squali del mare Mediterraneo*. Milano: Class Editori.

De Maddalena, A., Glaizot, O. and Oliver, G. (2003). On the great white shark, *Carcharodon carcharias* (Linnaeus, 1758), preserved in the Museum of Zoology in Lausanne. *Marine Life, 13*, (1/2), 53–59.

De Maddalena, A. and Hollà, A. (2006). Žraloky Stredozemného mora. *Enviromagazín*, July 2006: 28–29, 14.

De Maddalena, A., Preti, A. and Polansky, T. (2007). *A guide to the sharks of the Pacific Northwest (Including Oregon, Washington, British Columbia and Alaska)*. Madeira Park: Harbour Publishing, Madeira Park.

De Maddalena, A., Preti, A. and Smith, R. (2005). *Mako sharks*. Malabar: Krieger Publishing.

De Maddalena, A. and Révelart, A. L. (2008). *Le grand requin blanc sur les côtes françaises*. Hyéres, France: Turtle Prod Éditions / Média Plongée.

De Maddalena, A. and Zuffa, M. (2008). Historical and contemporary presence of the great white shark, *Carcharodon carcharias* (Linnaeus, 1758), along the Mediterranean coast of France. *Bollettino del Museo Civico di Storia Naturale di Venezia, 59*.

De Maddalena, A., Zuffa, M., Lipej, L. and Celona, A. (2001). An analysis of the photographic evidence of the largest great white sharks, *Carcharodon carcharias* (Linnaeus, 1758), captured in the Mediterranean Sea with considerations about the maximum size of the species. *Annales, Series historia naturalis, 11*, (2), 193–206.

DeMarte, R. (2007). Frank Mundus interview, August 2007. *Nor'East Saltwater*, September 5, 2007, 18, (23).

De Rosier, K. (2006). Melli, the model guy. *San Diego Natural History: Field Notes*, September 2006.

Digital Fish Library (2008): *Carcharodon carcharias* (White Shark). <http://www.digitalfishlibrary.org/library/>

Ellis, R. and McCosker, J. E. (1991). *Great white shark*. Stanford: Stanford University Press.

European Commission (2007). *Consultation on an EU action plan for sharks*. European Commission, Brussels.

Ezcurra, J. M., Mollet, H. F., O'Sullivan, J. B. and Lowe, C. G. (1996). *Captive feeding and growth of a juvenile white shark,* Carcharodon carcharias, *at the Monterey Bay Aquarium*. 22nd annual meeting of the American Elasmobrach Society in New Orleans, Louisiana, 1996.

Fergusson, I. K. (1996). Distribution and autecology of the white shark in the Eastern North Atlantic Ocean and the Mediterranean Sea. In A. P. Klimley and D. G. Ainley (Eds.), *Great white sharks: The biology of* Carcharodon carcharias (pp. 321–345). San Diego: Academic Press.

Ferreira, C. A. and Ferreira, T. P. (1996). Population dynamics of white sharks in South Africa. In A. P. Klimley and D. G. Ainley (Eds.), *Great white sharks: The biology of* Carcharodon carcharias (pp. 381–391). San Diego: Academic Press.

Fink, W. L., Hartel, K. E., Saul, W. G., Koon, E. M. and Wiley, E. O. (1979). *A report on current supplies and practices used in curation of Ichthyological collections. Ad Hoc Subcommittee on Curatorial Supplies and Practices of the ASIH Ichthyological Collection Committee*. American Society of Ichthyologists and Herpetologists, Washington, D.C.

Follett, W. I. (1966). Man-eater of the California coast. *Pacific Discovery, 19*, (1), 18–22.

———. (1974). Attacks by the White Shark, *Carcharodon carcharias* (Linnaeus, 1758), in Northern California. *California Fish and Game, 60*, (4), 192–198.

Francis, M. P. (1996). Observations on a pregnant white shark with a review of reproductive biology. In A. P. Klimley and D. G. Ainley (Eds.), *Great white sharks: The biology of* Carcharodon carcharias (pp. 157–172). San Diego: Academic Press.

Froese, R. and Pauly, D. (Eds.) (2008). *FishBase*. World Wide Web electronic publication. <www.fishbase.org> accessed 10/2008.

Gabriotti, V. and De Maddalena, A. (2004). Observations of an approach behaviour to a possible prey performed by some great white sharks, *Carcharodon carcharias* (Linnaeus, 1758), at the Neptune Islands, South Australia. *Bollettino del Museo Civico di Storia Naturale di Venezia, 55*, 151–157.

Galaz, T. and De Maddalena, A. (2004). On a great white shark, *Carcharodon carcharias* (Linnaeus, 1758), trapped in a tuna cage off Libya, Mediterranean Sea. *Annales, Series historia naturalis, 14*, (2), 159–164.

Gamberini, E. (1917). *Monografia marittima della Sicilia Nord-Orientale*. Principato, Messina.

Goldman, K. J., Anderson, S. D., McCosker, J. E. and Klimley, A. P. (1996). Temperature, swimming depth, and movements of a white shark at the South Farallon Islands, California. In A. P. Klimley and D. G. Ainley (Eds.), *Great white sharks: The biology of* Carcharodon carcharias (pp. 111–120). San Diego: Academic Press.

Gottfried, M. D., Compagno, L. J. V. and Bowman, S. C. (1996). Size and skeletal anatomy of the giant megatooth shark *Carcharodon megalodon*. In A. P. Klimley and D. G. Ainley (Eds.), *Great white sharks: The biology of* Carcharodon carcharias (pp. 55–66). San Diego: Academic Press.

Gruber, S. H. and Cohen, J. L. (1978). Visual system of the elasmobranchs: state of the art 1960–1975. In E. S. Hodgson and R. F. Mathewson (Eds.), *Sensory biology of sharks, skates and rays* (pp. 11–105). U.S. Office of Naval Research, Arlington.

Guitart-Manday, D. and Milera, J. F. (1974). El monstruo marino de Cojimar. *Mar y Pesca, 104,* 10–11.

Hairr, J. (2004). *Great White Sharks From Cape Hatteras To the Florida Keys.* John Hairr.

Hansson, H. G. (1998). *NEAT (North East Atlantic Taxa): South Scandinavian marine Plathelminthes Check-List, Internet Ed.* <http://www.tmbl.gu.se/libdb/taxon/neat_pdf/NEAT*Plathelmint.pdf.>

Hewitt, G. C. (1967). Some New Zealand parasitic Copepoda of the family Pandaridae. *New Zealand Journal of Marine & Freshwater Research, 1,* 180–264.

Hewitt, G. C. (1979). Eight species of parasitic Copepoda on a white shark. *New Zealand Journal of Marine & Freshwater Research* 13 (1), 171.

Hewitt, J. C. (1984). The great white shark in captivity: a history and prognosis. *AAZPA 1984 Annual Proceedings,* 317–324.

Hodgson, E. S. (1987). The shark's senses. In J. D. Stevens (Ed.), *Sharks* (pp. 76–83). Hong Kong: Intercontinental Publishing Corporation Limited.

Ichthyology Collection Database of the Academy of Natural Sciences in Philadelphia, (2008). <http://data.acnatsci.org/biodiversity_databases/fish.php>

Ichthyology Collection Database of the American Museum of Natural History, (2008). <http://research.amnh.org/vertzoo/>

Ichthyology Collection Database of the California Academy of Sciences, (2008). <http://research.calacademy.org/research/Ichthyology/collection/index.asp>

Ichthyology Collection Database of the Harvard University Museum of Comparative Zoology, (2008). <http://collections.mcz.harvard.edu/Fish/FishSearch.htm>

Ichthyology Collection Database of the Museum of Vertebrate Zoology in Berkeley, (2008). <http://mvzarctos.berkeley.edu>

Ichthyology Collection Database of the Scripps Institution of Oceanography, (2008). <http://collections.ucsd.edu/mv/index.cfm>

Ichthyology Collection Database of the Smithsonian Institution National Museum of Natural History, (2008). <http://nhb-acsmith2.si.edu/emuwebvzfishesweb/pages/nmnh/vz/DtlQueryFishes.php>

Ichthyology Collection Database of the University of Florida, (2008). <http://www.flmnh.ufl.edu/databases/fish/>

Ichthyology Collection Database of the Virginia Institute of Marine Science, (2008). <http://www2.vims.edu/fish_collection/>

Kabasakal, H. (2003). Historical records of the great white shark, *Carcharodon carcharias* (Linnaeus, 1758) (Lamniformes: Lamnidae), from the Sea of Marmara. *Annales, Series historia naturalis, 13,* (2), 173–180.

_____. (2008). Two new-born great white sharks, *Carcharodon carcharias* (Linnaeus, 1758) (Lamniformes; Lamnidae) from Turkish waters of north Aegean Sea. *Acta Adriatica.* In press.

Klimley, A. P. and Anderson, S. C. (1996). Residency patterns of white sharks at the South Farallon Islands, California. In A. P. Klimley and D. G. Ainley (Eds.), *Great white sharks: The biology of* Carcharodon carcharias (pp. 365–379). San Diego: Academic Press, San Diego.

Klimley, A. P., Le Boeuf, B. J., Cantara, K. M., Richert, J. E., Davis, S. F., Sommeran, S. V. and Kelly, J. T. (2001). The hunting strategy of white sharks (*Carcharodon carcharias*) near a seal colony. *Marine Biology, 138,* 617–636.

Klimley, A. P., Pyle, P. and Anderson, S. C. (1996). Tail slap and breach: agonistic displays among white sharks. In A. P. Klimley and D. G. Ainley (Eds.), *Great white sharks: The biology of* Carcharodon carcharias (pp. 241–256). San Diego: Academic Press.

_____. (1996). The behavior of white sharks and their pinniped prey during predatory attacks. In A. P. Klimley and D. G. Ainley (Eds.), *Great white sharks: The biology of* Carcharodon carcharias (175–191). San Diego: Academic Press.

Bibliography

Kohler, N. E., Casey, J. G. and Turner, P. A. (1996). NMFS Length-length and length-weight relationships for 13 shark species from the Western North Atlantic. *NOAA Technical Memorandum NMFS-NE* 110, 1–22.

Kreuzer, R. and Ahmed, R. (1978). *Shark utilization and marketing*. Rome: FAO.

Last, P. R. and Stevens, J. D. (1994). *Sharks and rays of Australia*. Australia: CSIRO.

Leim, A. H. and Scott, W. B. (1966). Fishes of the Atlantic coast of Canada. *Fisheries Research Board of Canada Bulletin, 155*, 1–485.

LeMier, E. H. (1951). Recent records of the great white shark, *Carcharodon carcharias*, on the Washington coast. *Copeia 1951* (3), 259.

Levine, M. (1996). Unprovoked attacks by white sharks off the South African coast. In A. P. Klimley and D. G. Ainley (Eds.), *Great white sharks: The biology of* Carcharodon carcharias (pp. 435–448). San Diego: Academic Press.

Lineaweaver, T. H. III and Backus, R. H. (1969). *The Natural History of Sharks*. Philadelphia: J. B. Lippincott Co.

Linnaeus, C. (1758). *Systema Naturae per regna tria naturae, secundum classes, ordines, genera, species, cum characteribus, differentiis, synonymis, locis. Editio decima, reformata*. Laurentius Salvius, Holmiae.

Lipej, L., De Maddalena, A. and Soldo, A. (2004). *Sharks of the Adriatic Sea*. Knjiznica Annales Majora, Koper.

Litvinov, F. F. and Laptikhovsky, V. V. (2005). Methods of investigations of shark heterodonty and dental formulae's variability with the blue shark, *Prionace glauca* taken as an example. *ICES CM 2005/N:27*.

Long, D. J. and Jones, R. E. (1996). White shark predation and scavenging on cetaceans in the Eastern North Pacific Ocean. In A. P. Klimley and D. G. Ainley (Eds.), *Great white sharks: The biology of* Carcharodon carcharias (pp. 293–307). San Diego: Academic Press.

Long, D. J., Hanni, K. D., Pyle, P., Roletto, J., Jones, R. E. and Bandar, R. (1996). White shark predation on four pinniped species in central California waters: geographic and temporal patterns inferred from wounded carcasses. In A. P. Klimley and D. G. Ainley (Eds.), *Great white sharks: The biology of* Carcharodon carcharias (pp. 263–274). San Diego: Academic Press.

Long, J. A. (1995). *The rise of fishes: 500 million years of Evolution*. Baltimore: John Hopkins University Press.

Low, F. H. (1963). *Fishing is for me*. New York: William Morrow & Company.

Maisey, J. G. (1987). Evolution of the shark. In J. D. Stevens (Ed.), *Sharks* (pp. 14–16). Hong Kong: Intercontinental Publishing Corporation Limited.

Malcolm, H., Bruce, B. D. and Stevens, J. D. (2001). *A review of the biology and status of white sharks in Australian waters*. CSIRO Marine Research, Hobart.

Martin, R. A. (1995). *Shark smart: the divers' guide to understanding shark behaviour*. Vancouver: Diving Naturalist Press.

_____. (2003). *Field guide to the great white shark*. ReefQuest Centre for Shark Research, Special Publication No. 1, Vancouver.

Martin, R. A. Hammerschlag, N., Collier, R. S. and Fallows, C. (2005). Predatory behaviour of white sharks (*Carcharodon carcharias*) at Seal Island, South Africa. *Journal of the Marine Biological Association of the United Kingdom, 85*, 1121–1135.

McCosker, J. E. (1987). The white shark, *Carcharodon carcharias*, has a warm stomach. *Copeia, 1*, 195–197.

_____. (1999). *The history of Steinhart Aquarium: A very fishy tale*. Virginia Beach: The Donning Company.

Michael, S. W. (1993). *Reef sharks and rays of the world*. Monterey: Sea Challengers, Monterey.

Miller, D. J. and Collier, R. S. (1980). Shark attacks in California and Oregon, 1926–1979. *California Fish and Game, 67*, (2), 76–104.

Mollet, H. F. (2008). White sharks (*Carcharodon carcharias*) in captivity. *On elasmobranch research around Monterey Bay*. Website: <http://homepage.mac.com/mollet/Cc/Cc_captive.html>

Mollet, H. F. and Cailliet, G. M. (1996). Using allometry to predict body mass from linear measurements of the White Shark. In A. P. Klimley and D. G. Ainley (Eds.), *Great white sharks: The biology of* Carcharodon carcharias (pp. 81–89). San Diego: Academic Press, San Diego.

Mollet, H. F., Cailliet, G. M., Klimley, A. P., Ebert, D. A., Testi, A. D. and Compagno, L. J. V. (1996). A review of length validation methods and protocols to measure large white sharks. In A. P. Klimley and D. G. Ainley (Eds.), *Great white sharks. The biology of* Carcharodon carcharias (pp. 91–108). San Diego: Academic Press.

Monterey Bay Aquarium (2008a). White Shark Project. Monterey Bay Aquarium website: <http://www.montereybayaquarium.org/cr/whiteshark.asp#exhibiting.>

Monterey Bay Aquarium (2008b). *2008 White Shark Project Press Kit*. Monterey Bay Aquarium.

Morey, G., Martínez, M., Massutí, E. and Moranta, J. (2003). The occurrence of white sharks, *Carcharodon carcharias*, around the Balearic Islands (western Mediterranean Sea). *Environmental Biology of Fishes*, 68, 425–432.

Müller, J. and Henle, F. G. J. (1838). *Systematische beschreibung der plagiostomen*. Berlin: Veit.

Mundus, F. (2000). The 3,427 lb White Shark: The Largest Fish On Rod and Reel. *Frank Mundus website*: <http://www.fmundus.com>

Orr, R. T. (1959). Sharks as enemies of sea otters. *Journal of Mammalogy*, 40, 617.

Paleontology Collection Database of the San Diego Natural History Museum (2008). <http://nhb-acsmith2.si.edu/emuwebvzfishesweb/pages/nmnh/vz/DtlQueryFishes.php>

Postel, E. (1958). Sur la presence de *Carcharodon carcharias* L. 1758 dans les eaux tunisiennes. *Bulletin du Muséum National d'Histoire Naturelle de Paris, Ser. 2*, 30, 342–344.

Powell, D. C. (2001). *A fascination for fish. Adventures of an underwater pioneer*. Berkeley: University of California Press / Monterey Bay Aquarium.

Pratt, H. L. (1996). Reproduction in the male white shark. In A. P. Klimley and D. G. Ainley (Eds.), *Great white sharks: The biology of Carcharodon carcharias* (pp. 131–138). San Diego: Academic Press.

Pratt, H. L., Jr., Casey, J. G. and Conklin, R. B. (1982). Observations on large white sharks, *Carcharodon carcharias*, off Long Island, New York. *Fishery Bulletin*, 80, (1), 153–156.

Pyle, P., Schramm, M. J., Keiper, C. and Anderson, S. D. (1999). Predation on a white shark (*Carcharodon carcharias*) by a killer whale (*Orcinus orca*) and a possible case of competitive displacement. *Marine Mammal Science*, 15, (2), 563–568.

Randall, J. E. (1973). Size of the great white shark (*Carcharodon*). *Science*, 181, (4095), 169–170.

_____. (1986). *Sharks of Arabia*. London: Immel Publishing Co.

_____. (1987). Refutation of lengths of 11.3, 9.0, and 6.4 m attributed to the white shark, *Carcharodon carcharias*. *California Fish and Game*, 73, (3), 163–168.

Ravazza, N. (2005). Il tonno fatato. *Cosedimare* <www.cosedimare.com> 14 Febbraio 2005.

Reidarson, T. H. and McBain, J. (1994). Hyperglycemia in two great white sharks. *IAAAM Newsletter*, 25, (4).

Roedel, P. M. and Ripley, W. E. (1950). California Sharks and Rays. *Fish Bulletin*, 75, 1–88.

Saïdi, B., Bradaï, M. N., Bouaïn, A., Guélorget, O. and Capapé, C. (2005). Capture of a pregnant female white shark, *Carcharodon carcharias* (Lamnidae) in the Gulf of Gabès (southern Tunisia, central Mediterranean) with comments on oophagy in sharks. *Cybium*, 29, (3), 303–307.

San Diego Natural History Museum (2006). Fossil Mysteries: Making of an Exhibition — The Megalodon Shark. San Diego Natural History Museum website: <http://www.sdnhm.org/exhibits/mystery/makingof/shark/>

Skomal, G. B. and Natanson, L. J. (2003). Age and growth of the blue shark, *Prionace glauca*, in the North Atlantic Ocean. *Fishery Bulletin*, 101, 627–639.

Smale, M. J. and Heemstra, P. C. (1997). First record of albinism in the great white shark, *Carcharodon carcharias* (Linnaeus, 1758). *South African Journal of Science*, 93, 243–245.

Smith, M. F. L., Marshall, A., Correia, J. P. and Rupp, J. (2004). Elasmobranch Transport Techniques and Equipment. In M. Smith, D. Warmolts, D. Thoney and R. Hueter (Eds.), *The Elasmobranch Husbandry Manual: Captive Care of Sharks, Rays and their Relatives* (pp. 105–131). Columbus: Ohio Biological Survey, Inc.

Smith, S. E., Au, D. W. and Show, C. (1998). Intrinsic rebound potentials of 26 species of Pacific sharks. *Marine & Freshwater Research*, 49, 663–678.

Springer, S. (1939). The great white shark (*Carcharodon carcharias*) in Florida waters. *Copeia*, 1939: 114–115.

Stephens, T. (2001). Great white shark attracts attention at Long Marine Lab. February 12, 2001. *UC Santa Cruz Currents Online*: <http://www.ucsc.edu/currents/00-01/02-12/shark.html>

Stevens, J. D. (1987). Shark biology. In J. D. Stevens (Ed.), *Sharks* (pp. 50–75). Hong Kong: Intercontinental Publishing Corporation Limited.

Stillwell, C. (1991). The ravenous mako. In S. H. Gruber (Ed.), Discovering sharks (pp. 77–88). *Underwater Naturalist, Bulletin American Littoral Society*, 19,(4)-20,(1).

Strong, W. R. Jr. (1996a). Repetitive aerial gaping: a thwart-induced behaviour in white sharks. In A. P. Klimley and D. G. Ainley (Eds.), *Great white sharks. The biology of Carcharodon carcharias* (pp. 207–215). San Diego: Academic Press.

_____. (1996b). Shape discrimination and visual predatory tactics. In A. P. Klimley and D. G. Ainley (Eds.), *Great white sharks: The biology of Carcharodon carcharias* (pp. 229–240). San Diego: Academic Press.

Strong, W. R. Jr., Bruce, B. D., Nelson, D. R. and Murphy, R. D. (1996). Population dynamics of white sharks in Spencer Gulf, South Australia. In A. P. Klimley and D. G. Ainley (Eds.), *Great white sharks: The biology of Carcharodon carcharias* (401–414). San Diego: Academic Press.

Sumner, R. B., Osburn, R. and Cole, L. J. (1913). A biological survey of the waters of Woods Hole and vicinity. Section 3. A catalogue of the Marine fauna. *Bulletin of the United States Bureau of Fisheries*, 31, 549–794.

Tanaka, S., Kitamura, T. and Nakano, H. (2002). Identification of shark species by SEM observation of denticle of shark fins. *Col. Vol. Sci. Pap. ICCAT*, 54, (4), 1386–1394.

Taylor, L. (1985). White Sharks in Hawaii: Historical and Contemporary Records. In G. Sibley, J. A. Seigel and C. C. Swift (Eds.), Biology of the white shark. A symposium (pp. 41–48). *Memoirs of the Southern California Academy of Sciences*, 9, 1–150.

Tortonese, E. (1956). *Fauna d'Italia vol.II. Leptocardia, Ciclostomata, Selachii*. Bologna: Calderini.

Tricas, T. C. and McCosker, J. E. (1984). Predatory behavior of the white shark (*Carcharodon carcharias*) with notes on its biology. *Proceedings of the California Academy of Sciences*, 43, (14), 221–238.

Uchida, S., Toda, M., Teshima, K. and Yano, K. (1996). Pregnant white sharks and full-term embryos from Japan. In A. P. Klimley and D. G. Ainley (Eds.), *Great white sharks: The biology of Carcharodon carcharias* (pp. 139–155). San Diego: Academic Press.

Vannuccini, S. (1999). Shark utilization, marketing and trade. *FAO Fisheries Technical Paper*, 389, 1–470.

Watts, S. (2001). *The end of the line?* San Francisco: San Francisco.

Welden, B. A., Cailliet, G. M. and Flegal, A. R. (1987). Comparison of radiometric with vertebral band age estimates in four California elasmobranchs. In R. C. Summerfelt and G. E. Hall (Eds.), *Age and growth of fish* (pp. 301–315). Ames: Iowa State University Press.

West, J. (1996). White shark attacks in Australian waters. In A. P. Klimley and D. G. Ainley (Eds.), *Great white sharks: The biology of Carcharodon carcharias* (pp. 449–455). San Diego: Academic Press.

Wildlife Conservation Society (2004). *White Shark* Carcharodon carcharias: *status and management challenges. Conclusions of the Workshop on Great White Shark Conservation Research*. New York: Wildlife Conservation Society.

Wintner, S. P. and Cliff, G. (1999). Age and growth determination of the white shark, *Carcharodon carcharias*, from the east coast of South Africa. *Fishery Bulletin*, 97, (1), 153–169.

Wood, F. G. (1959). Man eats maneater. *Mariner, Marineland of Florida*.

Zuffa, M., Van Grevelynghe, G., De Maddalena, A. and Storai, T. (2002). Records of the white shark, *Carcharodon carcharias* (Linnaeus, 1758), from the western Indian Ocean. *South African Journal of Science*, 98, (7–8), 347–349.

Index

Numbers in ***bold italics*** indicate pages with illustrations.

Abalone 54, 55, 130
ABC News 120
abundance 80–81
Academy of Natural Sciences 159–161
Achtheinus oblongus 68–69, 124
Acipenser medirostris see green sturgeon
adductor mandibulae ***40***, 41
age 31, 32, 44, 46–47, 52, 78, 122
agonistic behavior 66
Alaska State Museum 113
albacore 56, 182
Albatross 166
albinism 24
Algae 54, 55
Alligator 156, 157
Alopias sp. 57
Alopiidae 7, 29
American Museum of Natural History 154–156
Americo 137
amikacin sulfate 179
ampullae of Lorenzini 37
Ana Maria 118
anchovy 56, 118
angel shark 7
Animal (kingdom) 7
Anthosoma crassum 68
anus 34
apex predator 49–50
aplacental viviparity 44
Arctocephalus australis see South American fur seal
Arctocephalus forsteri see New Zealand fur seal
Arctocephalus pusillus doriferus see Australian fur seal
Arctocephalus pusillus pusillus see Cape fur seal
Arctocephalus townsendi see Guadalupe fur seal
Argyrosomus hololepidotus see Madagascar meagre
Ariidae 56
Arripis truttacea see Western Australia salmon
arteries 28–29, 41, 92
Asteris sp. 55

Atlantic bonito 56
Atlantic menhaden 56
Atomic Props 108, 111, 112
Atractoscion nobilis see white weakfish
atrium see auricle
attack site 62, 72, ***73***
attacks on humans 70–74
auricle 28–29
Australian fur seal 58
Australian sea lion 58
Austro-Hungarian Empire 74
autotrophs 50
Auxis rochei see bullet tuna
Auxis thazard see frigate tuna
axial skeleton 38

bacteria 96, 97
bai sa 9
bai sha 9
Balaenoptera musculus see blue whale
Balaenoptera physalus see fin whale
Barcelona Convention for the Protection of the Marine Environment and the Coastal Region of the Mediterranean 83
barracuda 56, 118, 179, 182
basal plate 25
basking shark 57, ***148***
bat eagle ray 57, 134, 135
Bay Eagle 163
beli morski volk 10
benthic 12, 25, 26, 27, 52, 54
Berkshire Museum 153–154
Bern Convention on the Conservation of European Wildlife and Natural Habitats 83
bijela ajkula 9
bile 34
bile duct 34
biodiversity 50
Birch Aquarium, Scripps Institution of Oceanography, University of California San Diego 100, 124, 126
bird 52, 54, 57, 180
bite 29, 30, 32, 44, 54, 62, 63, 64, 65, 66, 68, 72

bite-and-spit 62
bite scar 62, ***63***, 64–66, 132
bivalves 54, 55
Bivalvia 55
black rockfish 56, 136
bleach 96–97
blood 28–29, 34, 40, 41, 42, 52, 59, 62, 71, 92, 95, 174, 175, 179
blue pointer 9
blue shark 52, 57, 66, ***67***, 103, 170, 171
blue whale 58, 67
bluefin tuna see Northern bluefin tuna
bluefish 52, 56
body temperature 18, 29, 41
bone 7, 24, 37, 38, 92, 163
Bone Clones 79
bony fish 7, 9, 11, 24, 25, 26, 42, 52, 56, 69, 70, 74, 80
Bos sp. 58
bottlenose dolphin ***54***, 58
Bowman's capsule 42
brain 34, 35, ***35***, 38, 91, 140
braincase see chondrocranium
branchial arteries 28
breach 62, 63, 67
breeding areas 46
Brevoortia tyrannus see Atlantic menhaden
broadnose sevengill shark ***8***, 57, 178
bronze whaler 57
brown smooth-hound 57, 136
buccal cavity 26, 29, ***33***, 41
buffer 93, 142
bull ray 57
bull shark 55, 71, 89, 173
bullet tuna 56
bullhead shark 7
buoyancy 24–25
Bureau of Fisheries 99
Bursa californica see frog snail
butterfish 56
bycatch 74–77, 81, 172, 182

cabezon 56
cabosil 105
cage-diving 79

207

Index

calcium carbonate 93
California Academy of Sciences 30, 46, 88, 122, 128–139, 151, 171–172, 174–177, 179
California Department of Weights and Measures 115
California scorpionfish 179
California sea lion 58, 62, *63*, 118
California State University, Long Beach 124, 181
Callorhinus ursinus see northern fur seal 65
Callorhynchidae 57
camouflage pattern 23, 59
canavar köpekbaligi 10
cancer 78
Cancer antennarius see Pacific rock crab
Cancer magister see dungeness crab
Cancer productus see red rock crab
Canis lupus familiaris see dog
Cape cormorant 57
Cape fur seal 58, 62, 63, 64
capillaries 28, 42
Capra aegagrus hircus see domestic goat
captivity 28, 52, 87, 122, 132, 136, 168–186
Carangidae 56, 70
carbosil 98
carcharhinid 47
Carcharhiniformes 7
Carcharhinus brachyurus see bronze whaler
Carcharhinus leucas see bull shark
Carcharhinus obscurus see dusky shark
Carcharhinus plumbeus see sandbar shark
Carcharias atwoodi 8, 149, 150
Carcharias lamia 8
Carcharias maso 8
Carcharias rondeletti 8
Carcharias taurus see sandtiger shark
Carcharias verus 8
Carcharias vorax 8
Carcharodon 7, 8, 10, 11, 12, 31, 101
Carcharodon albimors 8
Carcharodon angustidens **10**
Carcharodon auriculatus **10**
Carcharodon capensis 8
Carcharodon megalodon see megatooth shark

Carcharodon orientalis 11
Carcharodon rondeletii 8
Carcharodon smithii 8
Carcharodon subauriculatus **10**
cardiac muscle 40, 41
cardiac stomach 33
cardinal veins
Caretta caretta see loggerhead sea turtle
carpet shark 7
cartilage 7, 25, 34, 37, 38, 39, 40, 78, 94, 95, 96, 97
cartilaginous fish 7, 37, 52, 57, 70, 77
cat 58
cattle 58
caudal peduncle 9, 12, 14, *17*, 18, 64, 65, 89, 90, 123
cell 7, 34, 36, 37, 42, 43, 89, 90
Center for Functional Magnetic Resonance Imaging 27, 33, 35, 36, 38, 39, 40, 45, 125–126
Center for Scientific Computation in Imaging 27, 33, 35, 36, 38, 39, 40, 45, 125–126
central nervous system 34, *35*
centrum 20, 39, 131, 136, 139, 141
ceratotrichia 40
cestode 68–69
cetacean 51, 54, 58, 65, 66
Cetorhinidae 7
Cetorhinus maximus see basking shark
Cheimerus nufar see santer seabream
Chelonia mydas see green sea turtle
chemical repellants 73
chemoreception 35
chimaera 7
chinook salmon 56, 135, 136
Chlamydoselachus anguineus see frilled shark
Chondrichthyes 7
chondrocranium 11, 34, *35*, 38, *38*, 39, 41, 87, 95, 150, 153, 160, *160*, 164
chondrocytes 7
Chordata 7
Chrysoblephus anglicus see English seabream
Chrysoblephus puniceus see slinger seabream
chub mackerel 56
circulatory system 28–29, 41, 92
clasper 42–44, *43*, 87, 117, 118, 135, 136, 141, 158
classification 7–8

clay 98, 104–107
Clistobothrium carcharodoni 69
cloaca 34, 42, 43, 44, 45
Coast Guard 130
coelom 43
collagen fiber 7
color 23–24
Commanche 115
common dentex 57
common dolphin 58
common names 9–10
common remora 69
competition 66–68
cone photoreceptor 37
Connecticut State Museum of Natural History 139, 140
connective tissue 7, 30, 37, 41, 94, 95, 96, 97
conservation 81–84
consumer 50
conus arteriosus 28
Convention on International Trade in Endangered Species of Wild Fauna and Flora 83
Convention on Migratory Species 83
copepod 68–69, *69*, 124
copulation 32, 42, 44
courtship 32, 44
cow shark 7
cranial nerve 34
Crickett II 141
crown 30
crustacean 32, 52, 55
cusp 25, 26, 30, 32, 46
cusplet 16, 32, **46**
cuttlefish 55
Cuvier's beaked whale 58
cystic duct 34

da bai sa 9
da bai sha 9
Dall's porpoise 58
Dasyatis sp. 57
decalcification 92, 93
Delphinus delphis see common dolphin
density 25
dental formula 31
Dentex dentex see common dentex
dentine 25, 30
Department of Marine Sciences, University of Connecticut 60, 99, 139–141
depth range 48–49
dermal denticle 11, 25–26, *26*, 30, 87, 127, 137, 141, 143

Index

Dermochelys coriacea see leatherback sea turtle
dexamethasone 179
digestion 29–34, 51
digestive system 29–34
Digital Fish Library 27, 33, 35, 36, 38, 39, 40, 45, 125–127
Dinemoura latifolia 68, 124
Dinemoura producta 68, 124
Dinobothrium septaria 69
disinfection 96–97
Distomum continuum 69
dog 58
dogfish shark 7
domestic goat 58
domestic pig 58
domestic sheep 58
dorsal aorta 28, 42
dried upper jaw perimeter 87
drying 89, 94–98, 99, 113–167
ductus efferens 42
dungeness crab 55, 165
duodenum 34
dusky dolphin 58
dusky shark 57
dusky smooth-hound 57, 159
dwarf sperm whale 58

ear 7, 35, 36, 37
earring 113, *113*
Echeneis naucrates see live sharksucker
echinoderm 55
Echthrogaleus coleoptratus 68, 124
Echthrogaleus denticulatus 68
ecotourism 79
eelgrass 55
egg case 44
elasmobranch 52, 69, 80
Elasmobranchii 7
elastin fibers 7
electric repellants 73
electroreception 35, 37
elephant seal see Northern elephant seal
embalming 89, 144–145, *146*
embryo *16*, 37, 40, 44, 45
enamel height of the largest upper tooth 87
endolymphatic ducts 36
English seabream 56
Engraulidae 56
Enhydra lutris see sea otter 65
enzymes 33, 34, 90, 92, 93
epaxial musculature 41
epidydimis 42
epipelagic zone 48, 77

equilibrium 36
Equus caballus see horse
Eschrichtius robustus see grey whale
esophagus 29, 32, 33
estimation of size 18–21
ethanol see ethyl alcohol
ethyl alcohol (ethanol) 90, 92, 93, 94, 120, 124, 129–137, 143, 146, 158, 159, 161, 163, 166
ethylene glycol 93
Etmopterus spinax see velvet belly
Eumetopias jubatus see Steller sea lion
European Community 83
European Habitats Directive 83
European pilchard 56
evolution 10–12
excretory system 34, 41–42
external anatomy 12–14, *12*, *13*
eye 14, 24, 26, 37, 87, 95, 99, 112, 143, 153, 164
eyeball 37

fecundity 45, 80
Felis catus see cat
fetus 44
fiberglass 99, 100, 104, 105, 107, 111, 139, 140, 141, 153, 154, 162, 171, 175, 176
Field Museum 146–148
fin whale 58, 67
finning 75, 81, 83, 84
fins 11, 9, *12*, 14, 17, 18, 19, 22, 24, 26, 28, 32, 34, 35, 39, 40, 41, 43, 48, 59, 60, 65, 67, 68, *69*, 73, 74, 75, 77, 78, 83, 87, 89, 91, 92, 93, 98, 99, 100, 103, 105, 107, 108, 111, 112, 115, 117, 118, 121, 122, 123, *123*, 124, 135, 141, 142, 143, 146, *147*, 150, 151, 155, 164, 169, 172, 183
Fish Market (restaurant) 103
fish trap 74
fisheries 74–77
Fishing Is for Me 155
fixation 89–90, 92–94, 98
Florida Museum of Natural History, University of Florida 101, 103, 104, 141–144
Florida Paleontological Supply 11
flounder 56, 174
Flying Fish 174
foam 92, 97, 98, 104–106, 108, 111

follicle 43
Food and Agriculture Organization of the United Nations 77
fork length 18–19, *19*
formaldehyde 89–91, 93
formalin 89–94, 98, 124, 136, 142–146, 163, 166
formic acid 90, 93
fossil 10–11, *10*, *11*, 31, 82, 108, 111
Fossil Mysteries 82, *108*, *109*, *110*, 111
freezing 89, 90, 103, 122, 123, 130, 161, 165, 166
frigate tuna 56
frilled shark 7, *8*
frog snail 55

Gadidae 56
Galeocerdo cuvier see tiger shark
Galeorhinus galeus see tope shark
gall bladder 34
game-fishing 75, 77
gannet 57
gas bladder 24–25
gaseous exchange 28
gastropod 54
Gastropoda 55
geographical distribution 47
gestation 45, 80
giant guitarfish 57
giant petrel 57
gill 26, 27, 28, 29, 33, 42, 177
gill arch *27*, 38, 39, 87, 100, 136, 172
gill chamber *27*, *36*, 41
gill lamella 26
gill net 74, 83, 114, 117, 118, 122, 123, 124, 125, 127, 128, 132, 135, 136, 137, 144, 145, 159, 165, 173, 176, 177, 180
gill slit 12, *12*, 14, 26, 32, 41, 90, 143, 170
give way 66
Global Shark Attack File 71
glomerulus 41–42
glycogen 41
goblin shark *8*
Grampus griseus see Risso's dolphin
grand requin blanc 9
grande squalo bianco 10
green sea turtle 57
green sturgeon 55
greenling 130
grey-headed gull 57
grey seal 58

209

Index

grey smooth-hound shark 57
grey whale 52, 58
ground shark 7
ground substance 7
growth rings 47
Guadalupe fur seal 58
gull 57
gustatory receptor 37

habitat 48–49
haemal arch 39, 40
hair cell 36, 37
hake 56
halibut 56, 123, 124, 132, 136, 144, 145, 171, 176, 177, 179, 182
Halichoerus grypus see grey seal
Haliotis sp. see abalone
hapuka 56
harbor porpoise 58
harbor seal 52, 53, 55, 57, 62, 165
harpoon 74, 75, 103, 137, 139, 162
Hawaii Marineland 169
hearing 36
heart 28–29, **28**, 41, 87, 92, 118, 139, 140, 142, 143, 157, 158
hepatic duct 34
Heterodontiformes 7
Heterodontus francisci see horn shark
heterotrophs 50
Hexagrammos sp. 130
Hexanchiformes 7
hidden approach 59
hohojirozame 10
Holokahana I 169
Homo sapiens sapiens see human
Honolulu Star Bulletin 169
horn shark **8**
horse 58
Houston Museum of Natural Science 162
human 45, 58, 59, 61, 62, 65, 68, 70–74, 80, 85
humpback whale 58
Hunter 131
hunting practice 65, 72
hvid haj 9
hvithai 10
hydrodynamic design 9, 12, 17–18
hydrogen peroxide 97
Hydrolagus colliei see spotted ratfish
hyoid arch **27**, 38, 39, 95, 160
hyomandibular 39, 95, 96

hyostylic suspension 39
hypaxial musculature 41
hyperglycemia 179

ileum 34
Imperial Maritime Austrian Government 74
Indo-Pacific humpbacked dolphin 58
insulin 179
intestinal valve 34, 94, 140
intestine 29, 34
isopropanol see isopropyl alcohol
isopropyl alcohol (isopropanol) 89, 90, 92, 93, 128, 130, 139, 143
Isurus 7, 14–16
Isurus oxyrinchus see shortfin mako
Isurus paucus see longfin mako
Italian Great White Shark Data Bank 85, 86

J&D Seafood 120
jack 56
jack mackerel 59
jackass penguin 57, 65
jaquetón blanco 10
jaw 11, 20, 29–32, **31**, 38–39, **40**, 41, 53, 63, 67, 73, 74, 75, 78, 79, 82, 86, 87, 89, 94–98, **96**, 100, 101, 103, 105, 107, 108, 114–167, **115**, **145**, **152**, **160**, **165**, 170, **170**, 171, 176
Jaws (movie) 5
joint 7, 38, 40, 95, 97
jump 18, 62, 67

kalb 9, 10
kàrkharos 8
keel 9, **12**, 14, 16, **17**, 18
kelb abjad 10
kelb el b'har 10
kelp 54, 55, 119
kidney 28, 41–42
killer whale 68
Kogia breviceps see pygmy sperm whale
Kogia simus see dwarf sperm whale

labial furrow 29
lagena 36
Lagenorhynchus obliquidens see Pacific white-sided dolphin
Lagenorhynchus obscurus see dusky dolphin

Lamna 7, 14–17
Lamna ditropis see salmon shark
Lamna nasus see porbeagle
Lamnidae 7, 9, 14, 15, 29
Lamniformes 7, 29
Laridae 57
Larus cirrocephalus see grey-headed gull
lateral septum 41
lateral snap 63
Latin name 8
leatherback sea turtle 57
lefkos karkarias 9
lemon shark 173
length-to-length relationships 18–20
length-to-weight relationships 18–20
lesser guitarfish 57
Letterman General Hospital 130
Life Through Time 136
ligament 39, 95, 96
ling 56
lingcod 56, 136, 178
litter size 45
Little Rose 139
live sharksucker 69, **70**
liver 25, 34, 42, 43, 74, 78, 117, 120, 139, 140, 165
loggerhead sea turtle 57
longfin mako 7, 14–15, **15**
longline 74, 76–77, 142, 143, 156, 159, 164
love bite 32, 44, 66
Lyons and O'Haver 99, 100, 103, 124, 176

mackerel shark 7, 9
Macronectes sp. 57
Macrosystis sp. 54
Madagascar meagre 56
magnetic resonance image **27**, **33**, **35**, **36**, **38**, **39**, **40**, 45, 125–126, **125**, **126**
mammal 52, 54, 58, 61, 65, 72
mandibular arch **27**, 38
mandibular cartilage 38
maneater 9
marele rechin alb 10
Maria 139
Marine Hall 114
marine mammal 52, 54, 61, 72
marine turtle see sea turtle
Marine Wildlife Veterinary Care and Research Center of the California Department of Fish and Game 123, 124
Marineland of Florida 169–170

Index

Marineland of the Pacific 169, 175
Marsea 142
mating 32, 43–44, 80
mating scar *see* love bite
maturity 42–44
maximum size 21–23, **23**
measurement 18–21, **19**, **119**
meat 74, 77–78, 82, 176
mechanoreception 35–36, 37
Meckel's cartilage *see* mandibular cartilage
Mediterranean monk seal 58
Megachasmidae 7
Megalodon: Largest Shark That Ever Lived **101**, **102**, 104
Megaptera novaeangliae see humpback whale
megatooth shark **10**, 11, **11**, **82**, 101, 104, **108**, **109**, **110**, 111, 112
mercury 78
Merluccius bilinearis see silver hake
Merluccius productus see North Pacific hake
meshing 72–73
Mesoplodon stejnegeri see Stejneger's beaked whale
metabolism 52, 168
Michigan State University Museum 111
migration 37, 48, 49
milk shark 57
Milstein Hall of Ocean Life 154, 155
Mirounga angustirostris see Northern elephant seal
Mitsukurina owstoni see goblin shark
Mitsukurinidae 7
mollusc 52, 55
Monachus monachus see Mediterranean monk seal
Monterey Bay Aquarium 28, 122, 168, **168**, 177–179, **178**, 181–186, **185**, **186**
Morone saxatilis see striped bass
Moss Landing Marine Laboratories 120–122, 123, 178
Mote Marine Laboratory 24, 145–146
Mote Marine Laboratory Aquarium 144–146
mouth 12, 26 27, 29, 31, 32, 33, 37, 38, 41, 44, 68, 90, 100, 104, 112, 119, 131, 153, 154, 157, 164, 170

movements 49
Ms Debb 142
Mugil sp. 57
mullet 57
muscular system 40–41
Musée cantonal de Zoologie de Lausanne 21, 22
Museum of Comparative Zoology, Museum of Natural History, Harvard University 69, 89, 149–153, 164
Museum of Discovery 105
Museum of Science 98, 148–149
Museum of Vertebrate Zoology, University of California 114
Museum of Zoology in Lausanne 21, 22
Mustelus californicus see grey smooth-hound shark
Mustelus canis see dusky smooth-hound
Mustelus henlei see brown smooth-hound
Mustelus sp. 32, 78
mutualism 69–70
Myliobatis californica see bat eagle ray
myoseptum 41
myotome 41
Mystic Marinelife Aquarium 99, 139

nasal flap 14, 35, 112
Natal pandora 56
Natal Sharks Board 72, 97
National Geographic 126
National Marine Fisheries Service 83, 141, 157
National Marine Fisheries Service Narragansett Laboratory 141, 157
National Ocean Service Marine Forensics Branch, Center for Coastal Environmental Health and Biomolecular Research 161–162
National Science Foundation 27, 33, 35, 36, 38, 39, 40, 45, 101, 104, 125–126
Natural History Museum, Humboldt State University 88, 129
Natural History Museum of Los Angeles County 31, 96, 114–120, 166
Natural History Museum of Trieste 74
natural selection 50
necropsy **121**

Negaprion brevirostris see lemon shark
Nemesis lamna 69
Neophoca cinerea see Australian sea lion
Neptunic *see* shark-proof suit
nervous system 34–37
Nesippus orientalis 68
neural arch 34, **36**, 39
neurocranium *see* chondrocranium
neuromast 35
New Zealand fur seal 58
New Zealand sea lion 58
newborn 14, 24, 32, 43, 89, 169
nictitating membrane 14, 37
North Carolina Maritime Museum 156–157, 159
North Carolina State Fair 157
North Carolina State Museum of Natural Sciences 28, 43, 91, 157, 158–159
North Museum of Natural History & Science 159
North Pacific hake 53, 56, 165
Northeast Fisheries Science Center 75, 76
northern bluefin tuna 56, 65, 75, 77, 182
northern elephant seal 52, 58, 62, 63, 64, 103, 118, 121
northern fur seal 65, 72
nostril 14, 35
notochord 40
Notorynchus cepedianus see broadnose sevengill shark
nursery area 45–46, 176

obligate ram ventilator 27, 176
Ocean Quest 176
ocean sunfish 182
odón 8
Odontaspididae 7
odontocete 65
olfaction 35
olfactory bulb 35
Oncorhynchus sp. 53, 56, 165
Oncorhynchus tshawytscha see chinook salmon
Ophiodon elongatus see lingcod
Orcinus orca see killer whale
Orectolobiformes 7
Orectolobus maculatus see spotted wobbegong
osmosis 42
ostium 43
otariid 62
Otariidae 52

Ottis' Fish Market 157
Outer Bay exhibit 182–186
ovary 43–44
overfishing 44, 74, 80
oviduct 43
oviparity 44
Ovis aries see domestic sheep
ovum 43–44
oxygen 26, 27, 28, 93, 170, 171, 174
oxytetracycline 47

Pacific rock crab 55, 136
Pacific salmon 56
Pacific white-sided dolphin 58
Pagrus auratus see squirefish
palatoquadrate cartilages 38
Paleocene 11
pancreas 34
pancreatic duct 34
Pandarus bicolor 68, 124
Pandarus satyrus 68
Pandarus smithii 68
papillae 33
Paralichthyidae 56
parallel swimming 66
parasite 68–69, *69*
parental care 46
Pargellus natalensis see Natal pandora
parturition 45
pas ljudožder 9
pectoral girdle 40
pelagic 12, 25, 26, 27, 52, 80, 119, 168, 170, 182
Pelagic Shark Research Foundation 1, 17, 50, 122–124, 181
pelican 57
Pelicanidae 57
pelvic girdle 40
pericardial cavity 28
peripheral nervous system 34
peristaltic pump 92
peritoneum 41
pescecane 10
peshkagen njeringrenes 9
pH 92
Phalacrocorax capensis see Cape cormorant
pharynx 26, 29, 32, 33, *33*, 37
philopatry 49, 68
Phoca vitulina see harbor seal
Phocarctos hookeri see New Zealand sea lion
Phocoena phocoena see harbor porpoise
Phocoenoides dalli see Dall's porpoise

photo-identification 49
photoreception 35, 36–37
photoreceptor 37
photosynthesis 48
Phyllobothrium loliginis 69
Physeter macrocephalus see sperm whale
pilchard 52
pinniped 51, *51*, 52, 54, 58, 60, 62, 63, 64, 65, 67, 68, 72, 163
Pioneer Fish Co. 118
Pisces 7
placenta 44
placental viviparity 44
placoderm 11
placoid scale *see* dermal denticle
plant 55
Platichthys stellatus 179
Pleuronectidae 56
plownose chimaera 57
Point Reyes Bird Observatory Conservation Science 181
polaris breach 62–63
Polyprion oxygeneios see hapuka
polyurethane 98
Pomadasys commersonni see smallspotted grunter
Pomatomus saltatrix see bluefish
popup archival tag 181, 182, 184, 186
porbeagle 7, 14–16, *15*, 74, 78
Porifera 55
porpoise 65, 99
precaudal length 18–19, *19*
precaudal pit 14, *17*, 18
predator 68
Predators (exhibition) **105**
predatory tactic 23, 59–65, 66
preservation in liquid 89–94
Prionace glauca see blue shark
Prionotus sp. 56
Pristiophoriformes 7
Procalo 118
producer 50
propulsion 17, 41
propylene phenoxcetol 93
proteoglycan 7
Protocol Concerning Specially Protected Areas and Biological Diversity in the Mediterranean 83
Pseudocarchariidae 7
Pteromylaeus vinus see bull ray
pterygiophore 40
pulp 25, 30
purse seine 74, 120, 163
pygmy sperm whale 58

pyloric stomach 33
pyramid of biomass 49, 50, 68

qarha levana 9

rate of food consumption 50–52
ray 7, 118, 172
Rebel 175
rechin alb 10
rectal gland 34
rectum 34
Red List of Threatened Species 83
red muscle 29, 40–41
red rock crab 55
regional endothermy 18, 29, 41
remora 69–70
Remora remora see common remora
Remorina albescens see white suckerfish
renal corpuscle 41
repetitive aerial gaping 68
replacement teeth 30, *30*, 94, 95, 97
replica 22, *60*, *79*, 87, *88*, 99–104, *99*, *100*, *101*, 124, 128, *129*, 139–141, *140*, *153*, *154*, 155, 156, 159, 162, 176
replica jaws *79*
reproduction 44–46
reproductive system 34, 41, 42–44
reptile 57
requiem shark 31
resin 98, 99, *106*, 107, 111, 140
respiratory system 26–28
rete mirabile 29, 41
retina 37
Rhincodon typus see whale shark
Rhinobatus annulatus see lesser guitarfish
rhipidion 43
Rhizoprionodon acutus see milk shark
Rhynchobatus djiddensis see giant guitarfish
ring valve 34
Risso's dolphin 58
rockfish 53, 56, 165, 178, 179
rod-and-reel 74
rod photoreceptor 37
root *see* basal plate
Roundabout 172, 174, 176
rugae 33

sacculus 36
salmon shark 7, 14–16, *15*, *16*

Index

San Diego Natural History Museum 82, 108, 111–112
sandbar shark 52, 57, 157, 160
sandtiger shark 57
Santa Maria 120
santer seabream 56
Sarda sarda see Atlantic bonito
Sardina pilchardus see European pilchard
Sardinops sagax see South American pilchard
saw shark 7
scalloped hammerhead **8**, 57, 182, 183
scavenger 50, 52, 61, 65, 66, 67, 74
School of Fisheries, University of Washington 165
Science Museum of Minnesota 108
scientific nomenclature 7–8
Scomber japonicus see chub mackerel
Scombridae 9, 75, 76
Scorpaena guttata see California scorpionfish
Scorpaena sp. 56
Scorpaenichthys marmoratus see cabezon
scorpionfish 56
Scripps Institution of Oceanography, University of California San Diego 27, 33, 35, 36, 38, 39, 40, 45, 69, 100, 124–128, 137, 151, 169
scroll valve 34
sculpture 89, 103, 104–112, **105, 106, 107, 108, 109, 110**
sea catfish 56
Sea Dancer 143
sea lion 52, 55, 59–64
sea otter 52, 59, 65, 72, 137
sea star 55
Sea Studios 178
sea turtle 51, 52, 54, 70
Sea World Aurora 103
Sea World Orlando 103
Sea World San Diego 162, 170–171, 172–174, 179
searobin 56
Sebastes melanops see black rockfish
Sebastes sp. 53, 56, 165, 178
Selachimorpha 7
seldevaja akula 10
semi-circulars canal 36
seminal vesicle 42
seminiferous tubule 42

senses 35–37
Sepiidae 55
shark cage 59, 73, 79, **105**
Shark Encounter 172, 173
shark fin soup 78
Shark Finning Prohibition Act 83
shark-proof suit 73
Shark Research Institute 71
Shark Week 126
Shax'dax' ooxu 113
Sheldon Jackson Museum 113
shell gland 43–44
Shima Marineland 172
shortfin mako 5, 7, **9**, 14–16, **15, 16**, 32, 52, 57, 74, 78, 170, 172, 173
silicone rubber 105
Silurian 11
silver hake 56
silver nitrate 46
sinus venosus 28–29
siphon sac 43
size 18–21
size at birth 45
skeletal muscle 40, 41
skeletal system 37–40
skin 25–26, **26**, 29, 35, 41, 78, 87, 89, 92, 95, 96, 97, 98, 99, 111, 116, 118, 119, 127, 129, 135, 136, 144, 145, 161
skin-mounting 89, 98, **148**
skull **38**, 94, 150
slinger seabream 56
smallspotted grunter 56
smart position-only tag 182, 183, 184
smell 35
Smithsonian Institution National Museum of Natural History 165–167
smooth-hound 32, 78
smooth muscle 40, 41
snail 55
snout 9, 14, 15, 18, 19, 29, 30, 35, 37, 100, 103, 111
social hierarchies 66
sonar 65
Sousa plumbea see Indo-Pacific humpbacked dolphin
South American fur seal 58
South American pilchard 56
South Florida Museum 11
specific gravity 25
speed 18, 41, 59, 60, 62
sperm 42–44
sperm sac 42
sperm whale 58

spermatophore 42–43
spermatozoa 42
Spheniscus demersus see jackass penguin; African penguin
Sphyraena sp. 179
Sphyraenidae 56
Sphyrna lewini see scalloped hammerhead
spinal chord **36**, 39
spinal cord 34, 36, 64
spinal nerve 34
spiny dogfish 57, 179
spiracle 14, 26, 27
spiral valve 34
splanchnocranium **27**, 38, **38**
sponge 55, 156
spotted ratfish 57
spotted wobbegong **8**
spy-hopping 65
squalene 78
Squaliformes 7
squalo bianco 10
Squalus acanthias see spiny dogfish
Squalus carcharias 7
Squalus lamia 8
Squalus lamia 8
Squalus vulgaris 8
Squatiniformes 7
squid 54, 55, 174, 179
squirefish 56
Standard Fisheries 133
Stanford University
starry flounder 179
Steinhart Aquarium, California Academy of Sciences 130, 132, 136, 171–172, 174–177
Steinhart White Shark Acquisition Team 174
Stejneger's beaked whale 58
Steller sea lion 58, 62
Stenella coeruleoalba see striped dolphin
stingray 57
stomach 29, 33, 34, 51, 94
striated muscle 40, 41
striped bass 56, 135, 178
striped dolphin 58
striped mullet 156
Stromateidae 56
suction disk **70**
Sula sp. 57
Sundowner 153, 164
superficial constrictor 41
surface lunge 63
Sus scrofa scrofa see domestic pig
swimming 17–18, 25, 27, 29,

Index

35, 40, 41, 51, 60, 61, 66, 67, 68, 172, 173, 174, 175, 177, 179, 180, 184
swordfish 56, 74, 75, 76, 78, 103, 148
symphysis 32, 38, 95, 97

tagging 5, 49, 180–186
Tagging of Pacific Predators 181
tail slap 67
tapetum lucidum 37
taste 37
tauró blanc 9
taxidermy 86, 89, 98–104, 140, **148**
Taxidermy Plus 99, 140
teeth 7, 8, 10–12, 14, 15, 16, 20, 22, 30–32, **30**, **31**, 33, 37, 46, 63, 74, 75, 78, 79, 82, 86, **86**, 87, 94, 95, 97, 99, 104, 108, 111, 113, **113**, 115, **115**, 129, 130, 131, 132, 133, 134, 135, 137, 140, 146, 147, 151, 152, 153, 154, 155, 157, 159, 160, 162, 163, 164, 165, 166, 167
testis 42, 87, 118
Tetrarhynchus megacephalus 69
Teuthoidea 55
threat display 67, 73
thresher shark 29, 57, 103
Thunnus alalunga see albacore
Thunnus albacares see yellowfin tuna
Thunnus thynnus see Northern bluefin tuna
tiburón blanco 10
tide 64
tiger shark 55, 70, 157
Tlingit 113
tooth bed 30, 94
tooth replacement *see* replacement teeth
tope shark 57, 118, 135, 183
topping-off 93
total length 18–19, **19**
touch 37
touch receptor 37
Trachurus declivis see jack mackerel
trammel net 74, 121
trawl 74
trematod 68–69
tubarâo branco 10
tuna cage 77
tuna farm 77
tuna trap 74–76
Tursiops truncatus see bottlenose dolphin
Two Sons 128

umbilical scar 45
United States Navy 99
University of California, Davis 114, 181
University of California, San Diego 125
University of Connecticut, Avery Point 140
University of Milan 85
University of Washington Fish Collection 16, 164–165
upper jaw perimeter 20, 87
upper jaw protrusion 29, 63, 67, 95
urea 42
ureter 42
urinary sinus 42
urine 42
urogenital papilla 42
urogenital sinus 42
urogenital system 41–44
Urophycis sp. 56
uterine wall 44
uterus 43–44
utilization 77–79
utriculus 36

vagina 44
valkohai 9
Van Sommeran, Sean R. 1, 17, 123, **123**, 124
veins 28–29, 41
velika bijela ajkula 10
velika bijela psina 9
velvet belly **8**
ventral aorta 28, 41
ventricle 28
vertebra 11, 34, **36**, 39, **40**, 46, 47, 87, 115–122, 127, 128, 133, 136–138, 141, 143, 172
vertebral centrum 20, 39, 131, 136, 139, 141
vertebral column 29, 38–42, **39**, 87, 89, 115, 117, 131, 133, 134, 135, 136, 137, 139, 142
Vertebrata 7
vertical approach 60, 61, 63, 64
Virginia Aquarium & Marine Science Center 164
Virginia Institute of Marine Science, College of William and Mary 94, 153, 162–164
visceral arch **27**, 38
vision 36–37
vitamin A 78
vithaj 10
vitrodentine 25, 30

wahsh 9
weißer hai 9
Western Australia salmon 56
whale shark 57, 103
White Angel 137
white death 9
white muscle 40, 41
white pointer 9
white suckerfish 69
white weakfish 56
witdoodshaai 9
witte haai 9
Working on a dream 4
World Conservation Union 83

Xiphias gladius see swordfish

yellowfin tuna 56, 182
yolk-sac 44

Zalophus californianus see California sea lion
żarłacz biały 10
Ziphius cavirostris see Cuvier's beaked whale
Zoostera sp. 55
žralok biely veľký 10
žralok bílý 9

www.ingramcontent.com/pod-product-compliance
Ingram Content Group UK Ltd.
Pitfield, Milton Keynes, MK11 3LW, UK
UKHW050529150426
5217IPUK00026B/1856